Sex Differences in Brain and Behavior

A subject collection from *Cold Spring Harbor Perspectives in Biology*

Sex Differences in Brain and Behavior

A subject collection from *Cold Spring Harbor Perspectives in Biology*

EDITED BY

Cynthia L. Jordan
Michigan State University

S. Marc Breedlove
Michigan State University

COLD SPRING HARBOR LABORATORY PRESS
Cold Spring Harbor, New York • www.cshlpress.org

Sex Differences in Brain and Behavior

A subject collection from *Cold Spring Harbor Perspectives in Biology*
Articles online at cshperspectives.org

Executive Editor	Richard Sever
Managing Editor	Maria Smit
Senior Project Manager	Barbara Acosta
Permissions Administrator	Carol Brown
Production Editor	Diane Schubach
Production Manager/Cover Designer	Denise Weiss
Publisher	John Inglis

Front cover artwork: "Brain-Freud" by Bruno Mallart © 2007. Image reprinted with express permission from the artist in conjunction with the David Goldman Agency.

Library of Congress Cataloging-in-Publication Data

Names: Jordan, Cynthia L., editor. | Breedlove, S. Marc, editor.
Title: Sex differences in brain and behavior : a subject collection from Cold Spring Harbor Perspectives in Biology / edited by Cynthia L. Jordan, Michigan State University and S. Marc Breedlove, Michigan State University.
Description: Cold Spring Harbor, New York : Cold Spring Harbor Laboratory Press, [2023] | Includes bibliographical references and index. |
Summary: "The extent to which there are differences between the sexes is an area of interest to physiologists, neuroscientists, and clinicians, as well as social scientists and the general public. This book examines recent research on the biological basis of sex differences, including differences in the brain, behavior, the immune system, and disease states"-- Provided by publisher.
Identifiers: LCCN 2022010320 (print) | LCCN 2022010321 (ebook) | ISBN 9781621823971 (cloth) | ISBN 9781621823988 (epub)
Subjects: LCSH: Sex differences. | Brain--Sex differences. | Human behavior--Physiological aspects.
Classification: LCC QP81.5 .S475 2022 (print) | LCC QP81.5 (ebook) | DDC 612.6--dc23/eng/20220401
LC record available at https://lccn.loc.gov/2022010320
LC ebook record available at https://lccn.loc.gov/2022010321

10 9 8 7 6 5 4 3 2 1

Contents

Preface

SEX DIFFERENCES IN BRAIN AND BEHAVIOR: An eddy of calm in the tempest? Has there ever been more turmoil over the issues of sex and gender than today? At the onset of civilization, men were so firmly in control, and so adamant about female submissiveness, that there was little open discussion about the role of women. Thanks to two centuries of feminism, sexist assumptions of the past have been so widely questioned that the trickle of grumbling has erupted in a waterfall of vigorous debate: the Me Too movement, the fight for female reproductive rights, the questioning of strictly binary gender identities, the transitioning of children away from the gender assigned at birth, the advisability of puberty blockers, whether special considerations for people who have a uterus is transphobic, and the inclusion of both sexes in medical trials and research, are examples of where our notions about sex and gender are in remarkable flux. Few would hazard a guess about how or when these issues will be resolved, if they ever are.

While open discussion and debate is crucial for a democracy, many questions about sex and gender have become increasingly polarized, amplified by "alternative facts," the rise of both authoritarianism and anti-secular sentiment, and a long-simmering frustration with the slow pace of civil rights efforts. Anyone asking whether children voicing gender dysphoria should always be encouraged to transition immediately can expect vitriol from both activists who immediately label any hesitation as transphobic, and fundamentalist religious forces declaring any transition a perversion of Nature. Even within the confines of academia, where it is supposedly permissible to consider controversial ideas under the protection of academic freedom, tenured professors endanger their positions if they ask whether all pedophiles actually abuse children, whether taking a "wait and see" approach to children with gender dysphoria is preferable to immediate transition, or whether social influences can spread gender dysphoria among young adults.

Within this maelstrom of sex and gender, the half-century or so of debate about whether the differences in the behavior of women and men are due to differences in brain structure and/or function, whether circulating hormones contribute to those differences, and whether prenatal hormones help engender them, may seem staid and decorous in comparison. The chapters in this volume offer a variety of windows onto these more circumscribed issues while maintaining respect for divergent views and those who hold them.

For example, philosopher Cordelia Fine has a long history of skepticism about sex differences in the human brain and their significance for behavior, pointing out instances where these may have been exaggerated in an enterprise that incentivizes reporting of differences between groups rather than null findings (Fine 2017). In her chapter herein, she explicitly addresses the thorny issue of how to find the boundary when such critiques leave the realm of fair skepticism and cross into politicized discourse. Naturally, scientists who search for and report sex differences in the brain, versus those concerned that the significance of such reports are inflated by the public, may differ in their perception of where the boundary between reasonable skepticism and politically motivated denial falls. Objects in the mirror may be closer than they appear. Fine's reasoned analysis of this question in three domains offers an important framework for both sides in the future.

That framework will only gather in importance if more, and more subtle, sex differences in behavior come to light, as Brian Trainor and Annegret Falkner predict will occur with the advent of machine-learning tools and "big data," which are on the horizon. They argue that these approaches will provide a richer understanding of individual differences in behavior, both between and *within* the sexes, as opposed to traditional "single-variable" measures of complex behavior, which are typically built upon assumptions that what males do represents a default. That approach can mask individual variability within the constraints of male behavior. Trying to squeeze the behavior of females into a box predicated on male behavior may not only gloss over important subtleties, but may also prevent researchers from understanding the adaptive significance of alternative behaviors.

That largely implicit consideration of males as the default has a long history in the biomedical literature, as Irving Zucker et al. point out, with the result that most of the accepted notions about the kinetics of drug treatment, based as they are on measures from only males, again tries to squeeze females into a male-biased box, exposing women to higher drug concentrations and more adverse reactions than males. The authors document how, despite the NIH mandate to include both sexes in research and trials, research from the past decade continues to neglect females.

Perhaps the most well-known overmedication caused by ignoring sex differences in pharmacokinetics is the excess dosage of the so-called Z-drugs as a sleep aid in women. James Walton et al. note that another important variable typically ignored in the conduct of research is the circadian rhythm, which interacts with sex, as they detail. The sex differences in chronotype, which begin at adolescence in humans, are in full display in animal studies where circulating gonadal hormones, rather than sex chromosomes, appear to be responsible. Lumping measurements from different phases of the circadian rhythm not only adds to variance that might obscure differences, but can again mask important behavioral adaptations.

Our current understanding of the so-called organizational influence of gonadal hormones on brain structure and function is expanded by Bruno Gegenhuber and Jessica Tollkuhn's exciting recent discovery that estrogen receptors can induce a long-lasting sex bias in gene expression in the bed nucleus of the stria terminalis, with implications for other brain regions involved in social behaviors, including sexual and parental behavior. They map out a program for how to explore which regions display similarly sustained epigenetic changes, and how those changes interact to affect behavior.

One of the most prominent and persistent sex differences in human disorders is the greater risk of major depressive disorder (MDD) in women than in men. Elizabeth Williams et al. probe the roots of this sex difference in preclinical studies of rodents, focusing on a particular brain circuit, from the ventral hippocampus to the nucleus accumbens, where sex differences in excitability appear to underlie sex differences in behaviors, offering a strong parallel to sex differences in human MDD. They speculate that a better understanding of sex differences in this circuit could offer improved diagnosis, prevention, and treatment of depression and other mood disorders.

An important factor in depression is stress and, despite the trope that only death and taxes are certain, we can surely add stress as another inevitability of life. Here again, sex differences are prominent, perhaps because of sex differences in reproductive strategy and, especially in mammals, sex differences in parental investment. As Robert Handa and colleagues explain, the activation of the hypothalamic-pituitary-adrenal axis (HPA) in response to physical stressors is generally greater in females than males among mammals, but in humans, where stress tends to be psychological—at least in laboratory studies—men show greater activation of the HPA. They also relate recent explorations suggesting that gender identity is another factor in responding to stress, and call for studies of the interaction of chromosomal sex and gender identity to better understand the stress response.

Another crucial response to stress involves the innate immune system, including the mast cells that orchestrate defenses, but can sometimes overreact, with consequences for many inflammatory and stress-associated disorders, including autoimmune diseases and migraines. Emily Mackey and Adam Moeser lay out preclinical research that begins to parse the role of sex differences in mast cell response, which could underlie the prominent sex differences seen in many of these disorders. They find strong evidence that circulating gonadal steroids modulate mast cell responses, and that these same hormones exert a longer-lasting effect in the perinatal period.

Evidence for the influence of gonadal hormones in the perinatal period on human sexual orientation and gender-typical behavior is explored by Ashlyn Swift-Gallant et al., exploiting indirect markers for prenatal androgen influence—digit ratios—and a clinical syndrome in which prenatal hormone levels are sex-typical in early fetal development, but largely absent in the period just before and after birth. Together, these data offer evidence that androgen exposure in the early fetal period can promote sexual attraction to females in adulthood, while androgen levels in the late perinatal period affect childhood gender conformity, and estrogens in the later perinatal period may augment attraction to men in adulthood.

Because most of the sex differences in behavior common in mammalian species, such as in aggression, dominance, and territorial defense, are reversed in the spotted hyena, these outliers offer a fascinating insight into the evolutionary forces driving those sex differences. S. McCormick et al. provide an overview of the literature about these beautiful, fascinating, and ferocious animals, arguing that a critical distinction for understanding the evolution of the sex reversals is between behaviors related to *access* to food, rather than those related to *acquiring* food.

Arthur Arnold concludes the volume by calling for a greater appreciation for the role of chromosomal sex in the training of future scientists exploring sex differences in brain and behavior. Explicating the influence of mammalian sex chromosomes, and their fascinating evolutionary history, as well as the alternative systems of sex determination in other vertebrates, will provide a richer understanding of sexual differentiation in all its domains, and better prepare future researchers.

Since the onset of this book project, we have lost two outstanding scientists who were also devoted mentors and wonderful people. A student of Donald Hebb, Stephen E. Glickman never lost his sense of wonder about animal behavior, studying animals in zoos and an astonishing variety of mammals, including skunks, gerbils, lizards, wood rats, and those amazing, gender-bending spotted hyenas. Heading up a massive project to capture hyena pups in Kenya to create a pack roaming a semi-naturalistic compound in the hills of Berkeley, California, Steve was perhaps the only person who was so fluent in the languages of psychologists, endocrinologists, evolutionary biologists, and field ecologists to successfully coordinate this massive project that informed many of the discoveries reported in McCormick et al.'s chapter herein, for which Steve was a coauthor. Steve was also an excellent colleague for both the editors in their years at Berkeley. A gentle, self-deprecating man with a wonderful smile and a truly vast breadth of intellectual interests, he passed away unexpectedly at 87.

We also shared a personal history with Robert J. Handa, who was a graduate student in anatomy while we were psychology graduate students at UCLA. Bob was a keen observer and critical thinker who realized that studies relying upon 5-α reduced metabolites of testosterone such as dihydrotestosterone (DHT) to stimulate only androgen receptors were working from a flawed assumption. Through his careful analyses, Bob documented metabolic pathways that could convert DHT to ligands active at estrogen receptors, providing a vital corrective to the literature (unfortunately, not everyone got the memo). We fondly recall his grace and ready smile in every situation, and mourn his passing much too soon at the age of 67. In addition to his prolific research career, Bob was an avid fisherman and the ultimate mentor. On his final fishing trip, fully aware of his grave prognosis, Bob was pleased that his contribution to this volume, the last manuscript he completed, would help his students' careers. Would that we all had such considerate and supportive mentors, and may all who follow after us be so blessed!

We thank Richard Sever for inspiring this volume and Barbara Acosta for her guidance, patience, and good humor—despite the pandemic—and all our authors for their dedication to a better understanding of what it means to be a human of any gender.

CYNTHIA JORDAN
S. MARC BREEDLOVE

REFERENCE

Fine C. 2017. *Testosterone Rex: myths of sex, science, and society.* W.W. Norton, New York.

Fairly Criticized, or Politicized? Conflicts in the Neuroscience of Sex Differences in the Human Brain

Cordelia Fine

History & Philosophy of Science Programme, School of Historical & Philosophical Studies, The University of Melbourne, Victoria 3010, Australia

Correspondence: cfine@unimelb.edu.au

Investigations of sex differences in the human brain take place on politically sensitive terrain. While some scholars express concern that gendered biases and stereotypes remain embedded in scientific research, others are alarmed about the politicization of science. To help better understand these debates, this review sets out three kinds of conflicts that can arise in the neuroscience of sex differences: academic freedom versus gender equality; frameworks, background assumptions, and dominant methodologies; and inductive risk and social values. The boundaries between fair criticism and politicization are explored for each kind of conflict, pointing to ways in which the academic community can facilitate fair criticism while protecting against politicization.

Neuroscientific investigations of sex differences take place on politically sensitive terrain. Historically, this work has played a role in justifying restrictions on women's political rights and educational opportunities (Russett 1989; Schiebinger 1996) and, more recently, casting doubt on women's fitness for positions of power and responsibility (Fausto-Sterling 1992; Tavris 1993). Today, feminist perspectives on neuroscience argue that gendered biases and stereotypes often remain embedded in scientists' theories, hypotheses, methods, and interpretations, with potentially detrimental social effects (e.g., Jordan-Young 2011; Bluhm et al. 2012; Fine 2013; Bluhm 2020). This critical feminist work is sometimes published in mainstream (neuro)science journals (e.g., Eliot 2011; Rippon et al. 2014; Richardson et al. 2015; Eliot and Richardson 2016; Joel and Fausto-Sterling 2016), indicating an openness to such perspectives on the part of the mainstream neuroscience community. However, it is also sometimes claimed that such work (including, in the interests of disclosure, my own) politicizes science: for example, by accepting or rejecting theories or findings largely on the basis of their political palatability rather than the evidence (Baron-Cohen 2010; Cahill 2014; Greenberg et al. 2018), or that studying sex differences in the brain is politically dangerous, and discouraged, territory (Cahill 2017).

To help onlookers and participants parse these debates, and encourage more productive ones, I set out three distinct, though potentially overlapping, conflicts that can arise in the neuroscience of sex differences (Fig. 1). I largely focus here on conflicts that arise due to feminist criticism of this body of research, rather than between competing research approaches to the same phenomenon (e.g., van Anders 2013). The first concerns conflicts between academic freedom and gender equality. The second concerns the frameworks, background assumptions, and dominant methodologies used by scientists in the production of scientific knowledge. The third potential kind of conflict is over standards of evidence for making politically sensitive empirical claims, in light of the inevitability of the possibility of error.

In what follows, I provide examples of each kind of conflict, and offer suggestions for distinguishing fair criticism from politicization. I conclude with some practical suggestions for encouraging more of the former and less of the latter in this controversial domain of research.

CONFLICT 1: ACADEMIC FREEDOM VERSUS EQUALITY

Academic freedom is rightly a core value of universities. Academics' role as independent experts and trained critical thinkers within a community of scholars uniquely positions them to challenge prevailing religious, state, or popular orthodoxies in the pursuit of knowledge (for a recent account, see Stone and Evans 2021). To this end, academic freedom bestows scholars with the liberty to investigate, propose, and discuss unpopular and even offensive lines of inquiry or argument, albeit with the responsibility to do so while upholding disciplinary standards (and, of course, remaining within the bounds of the law). Tolerance for controversial ideas is critical not just to enable academics to "speak truth to power," but for the knowledge production process itself. As one defense of academic freedom puts it: a "knowledge claim gains objectivity and warrant to the degree that it is the product of exposure to the fullest range of criticisms and perspectives" (Anderson 1995, p. 198).

However, as a society, we do not value the knowledge that academic freedom facilitates so highly that this trumps all other values (Douglas 2009). The most familiar example of this point for neuroscientists is that scientists must submit to the requirements of research ethics committees, which place constraints on the kinds of methods that can be used to gain knowledge. Although specific research ethics principles and their application can be contested (e.g.,

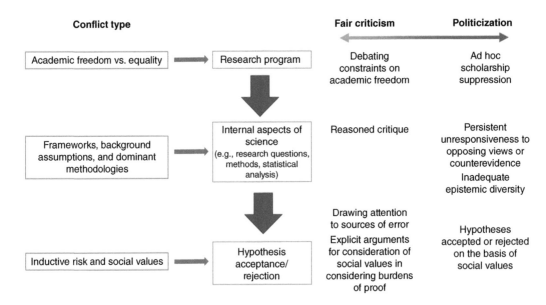

Figure 1. Fair criticism and politicization in three kinds of conflict over scientific research.

regarding what should count as the acceptable use of experimental animals, or the use of intentional deception with human participants), the idea that scientific methods should operate within the constraints of our general moral responsibilities to respect the rights of others is uncontroversial. Other kinds of constraints on research can be understood as likewise meeting those general obligations. For example, some lines of inquiry may require strict biosafety protocols (e.g., gain of function viral research) for the protection of public health. There are also sometimes proscriptions at the boundaries of the acquisition of knowledge and its use (e.g., the intentional creation of a human clone).

In addition, specific research ethics guidelines or policies have sometimes been developed in an attempt to redress group inequalities in the benefits gained from research. These are perhaps best understood not as outright suppression of particular lines of inquiry, but as influential changes in the "conditions that shape what scientists do, encouraging them to follow certain lines of inquiry and discouraging them from following others" (Johnson 1996, p. 208). In Australia, for example, in recognition of an unjust distribution of the burdens and benefits of research focused on Aboriginal and Torres Strait Islander communities (Bainbridge et al. 2015), the government's National Health and Medical Research Council (National Health and Medical Research Council (Australia) 2018a) published human research ethical guidelines specifically for research with these communities. The guidelines stress the importance of members of these communities (potentially including non-academics) playing an active role in shaping various aspects of research. For example, a companion document to the guidelines cites community involvement as potentially including activities such as: "give feedback into appropriateness of the research agenda and methodology"; "Request different approaches if required"; and "Check that there is agreement about the information in the reports" (National Health and Medical Research Council (Australia) 2018b).

Another, likely familiar, example is the U.S. National Institutes of Health (NIH) Revitalization Act (1993), which mandates the equal in-clusion of women and minority groups in publicly funded biomedical research. More recently, the NIH developed a similar mandate for basic and preclinical biomedical research, with the expectation that "sex as a biological variable will be factored into research designs, analyses, and reporting in vertebrate animal and human studies" (NIH 2015). Kourany (2016, p. 782) has argued that the NIH Revitalization Act constitutes "a constraint on scientists' freedom to design their own research programs, and it was justified by women's and minority men's right to equality—in this case their right to equality of access to health care." Although this dictate only applies to research that receives government funding, practically speaking, this will impose a significant constraint on scientists who would normally rely on such funding. It also sets a powerful norm. Indeed, in 2017, the *Journal of Neuroscience Research* announced a default requirement that authors "ensure proper consideration of sex as a biological variable" in submitted manuscripts to promote gender equality in the benefits of research (Prager 2017, p. 95). This includes the prescription that papers using subjects of only one sex must state this in the title and abstract and offer a rationale for this decision.

Some have been critical of such constraints on scientists' research—arguing, for example, that the 2015 NIH policy will not, in fact, create more equitable health outcomes between women and men (Joel et al. 2015; Richardson et al. 2015). For instance, Richardson et al. (2015, p. 13420) have argued that producing such knowledge requires "rigorous validation of animal models for the study of any particular human biological sex-difference pathway" and careful consideration of many gender- and sex-linked variables, some of which are uniquely human. Importantly, the authors do not object to investigations of sex effects in the brain, per se. But they argue that "policies mandating the study of sex-related variables in cells, tissues, and animal models are an impoverished approach to this issue." Indeed, they suggest that the policy may actively undermine its stated aim, being "more likely to introduce conceptual and empirical problems in research on sex and gender than bring new clarity to differences in men's and women's health outcomes."

Far more controversial is the idea that certain lines of inquiry should be forbidden because of the detrimental consequences for a particular marginalized group. Indeed, exploration of such arguments are largely circumscribed to the philosophical literature. Moreover, contrary to the idea that inherent group-based differences are a taboo topic, these arguments have been made on the grounds of the likelihood of continuing error and/or the opportunity costs of such research, rather than harms from knowledge per se. For example, Kourany (2016) has argued that women's and racial minorities' access to equality has been undermined by centuries of research investigating biologically based sex- and race-linked cognitive abilities. Even though this research invariably comes to be "contested and corrected" (p. 781), the process causes significant harm in its perpetuation of harmful stereotypes. In addition, she argues that such research creates an opportunity cost in terms of limited resources that might otherwise be spent on producing knowledge beneficial to women and racial minorities. While recognizing the costs of curtailment of academic freedom, Kourany nonetheless proposes "tighter restrictions on race- and gender-related cognitive differences research" (p. 779) on the general grounds that "scientists' right to freedom of research cannot be allowed to subvert other people's rights" (p. 789). Kourany (p. 786) does not propose an outright ban, but recommends constraints and oversights akin to those developed by the National Research Council (2004) in response to the recognition of terrorist threats from certain kinds of biotechnology research:

[N]ew policy constraints that include research guidelines for the weighing of societal harms of research against societal benefits, that include educational programs and codes of ethics designed to foster a culture of responsibility among social scientists, and that include a new National Science Advisory Board for Social Research … [that] could count among its members both leading social scientists and representatives of the public at large, such as racial, ethnic, and gender group advocates.

The philosopher Kitcher (2001) has put forward a somewhat related view. He draws on prima facie concerns that, in a society in which certain groups (such as women and members of racial minorities) are systematically disadvantaged, scientific research is likely to be inadvertently biased by sexist and racist values (a political asymmetry). Moreover, members of society (and even well-intentioned investigators) will tend to put greater credence in findings that reinforce stereotypes about social disadvantaged groups than in those that challenge them (an epistemic asymmetry). This, in turn, could result in reduced support for policies designed to promote these marginalized groups, while also having harmful psychological effects on members of those groups. Drawing on John Stuart Mill's famous defense of free expression (Mill 1859), Kitcher (2001, p. 95) argues that careful consideration of the underlying value of freedom of inquiry, if supplemented with evidence of political and epistemic asymmetries, allows for arguments in favor of a ban on such research:

To take seriously Mill's point that the freedom to which we aspire is the freedom to define and pursue our own vision of the good is to recognize the possibility that the unconstrained pursuit of inquiry might sometimes interfere with the most important kind of freedom, at least for some members of society. So we can envisage a Millian argument against freedom of inquiry, one that proceeds by trying to show that certain types of research would be likely to undermine a more fundamental freedom.

However, Kitcher ultimately concludes that even so, in practical terms, such restrictions would only exacerbate, rather than ameliorate, the underlying social issues. In particular, he anticipates it would lead to the proscription being misunderstood as a situation in which "everyone knows what the research would show and that people are unwilling to face the unpleasant truth" (p. 105).

Fair Criticism versus Politicization

The examples outlined above are not presented as endorsements of arguments justifying constraints on freedom of inquiry into sex differences. Discussing the many possible objections to such positions is beyond both the scope and purpose of this article. The point is simply that arguments debating constraints on academic freedom in the interests of gender (or other group-based) equality are as legitimate as arguments debating constraints on research in the interests of other social values (e.g., the voluntary moratorium on some recombinant DNA experiments established at the Asilomar conference in 1975 [Barinaga 2000]). Reasoned arguments for and against such constraints constitute fair criticism of lines of scientific inquiry, and scientists (and scholars from cognate disciplines) surely have distinct and valuable perspectives on these matters. Such commentary is also, itself, protected by academic freedom.

What, then, represents politicization in such conflicts? Even those willing to contemplate constraints on academic freedom in the interests of gender equality would surely agree that a baseline requisite would be comprehensive and transparent consultation with a broad range of stakeholders, in which the potential costs, as well as benefits, of constraints of varying degrees were thoroughly explored. Moreover, as with existing restrictions on academic freedom, any such constraints that were developed would have to be transparently articulated, and universally and consistently applied.

In contrast, ad hoc attempts to suppress particular lines of research, or the publication or presentation of particular findings by individuals or groups, represent unacceptable forms of politicization. Stevens et al. (2020) have identified three such sources of scholarship suppression in academia: external actors (e.g., interest groups, government); internal actors (students, university staff, and administrators); and a hybrid form in which external pressures lead to suppression by internal actors. Tactics used by economic, political, or emotionally vested interests include harassment or denunciation of scientists or their institutions, attempts to block research, delay or prevent publication, and to block spoken or written dissemination of findings (see Lilienfeld 2002; Rosenstock and Lee 2002; Stevens et al. 2020). Stevens et al. (2020) describe, for example, a typical chain of events in

which a paper, op-ed, or talk triggers an organized campaign or petition for the academic to be sanctioned, disciplined, or de-platformed.

It is unclear that such dynamics take place with regard to the study of sex differences in the brain. This line of inquiry has been characterized as "a terrific way for a brain scientist not studying reproductive functions to lose credibility at best, and at worst, become a pariah in the eyes of the neuroscience mainstream" (Cahill 2017, p. 12; see also Whipple 2016). However, an estimate of relative counts of publications relating to sex differences in the human brain (excluding medical research), found that these have "generally grown as a share of all publications since 1950" (Lockhart 2020, p. 371). Moreover, to my knowledge, there are no documented examples of actual or attempted scholarship suppression of neuroscientific investigations of sex differences in recent decades. However, there are a growing number of examples in closely related research areas, particularly relating to gender identity (e.g., Fazackerley 2018; Bailey 2019; Singal 2020; Cummins 2021; Suissa and Sullivan 2021).

CONFLICT 2: FRAMEWORKS, BACKGROUND ASSUMPTIONS, AND DOMINANT METHODOLOGIES

Scientists make many decisions in the course of doing research: from what hypotheses to test, to choices about methods, samples, measurement of the phenomena of interest, data characterization, analysis, interpretation of data, presentation of results, and so on. These choices are shaped by theoretical frameworks, background assumptions, and dominant methodologies. Such decisions are therefore often a rich source of conflicts, in all areas of science. The scientific study of sex differences is no exception, and an important goal of feminist perspectives on science has been to document "how gendered practices or assumptions in a scientific field prevented researchers from accurately interpreting data, caused inferential leaps, blocked the consideration of alternative hypotheses, overdetermined theory choice, or biased descriptive language" (Richardson 2010, p. 346).

Cite this article as *Cold Spring Harb Perspect Biol* doi: 10.1101/cshperspect.a039115

For example, one major target of feminist critique has been "brain organization theory" (e.g., Jordan-Young 2011; see also Bluhm 2020). Although now superseded by a more complex account (see McCarthy and Arnold 2011), this traditionally proposed a linear developmental pathway from genetic sex to gonadal hormones that, in turn, determine brain sex in the form of permanent structural and functional sex differences in the brain.

Whereas traditionally applied to the study of reproductive and mating behavior in nonhuman animals, brain organization theory has also shaped research questions, background assumptions, and dominant methodologies involved in investigating early hormonal influences on human gender identity, sexuality, and gendered interests. Consider, for example, investigations of girls with (classical) congenital adrenal hyperplasia (CAH), a population who, due to near male-typical levels of androgens in utero, are born with atypical or masculinized genitalia (White and Speiser 2000). Studies reliably find that girls with CAH show greater interest in male-typical toys, activities, and careers than control girls (e.g., Hines 2015; Spencer et al. 2021). These findings have typically been interpreted as direct effects of early brain masculinization (see Jordan-Young 2012). Thus, one background assumption of such research is that a linear model for relatively stereotypic nonhuman animal mating behavior is an appropriate one for understanding complex, intentional, and culturally and historically variable gendered human behavior. Another (tacit) background assumption is that the activities and careers designated "masculine" by researchers are universally and timelessly so. Both assumptions have long been the target of feminist critiques (e.g., Longino and Doell 1983; Bleier 1984; Doell and Longino 1988; Jordan-Young 2011; Grossi and Fine 2012).

These background assumptions, in turn, have shaped the dominant methodology in this area of research. For example, researchers have not attempted to categorize or create toys on the basis of the presence or absence of features or affordances explicitly theorized to be timelessly or universally appealing to a masculinized brain

(Fine 2015), such as movement, color, and form (Alexander 2003), affordance for active play (Alexander and Saenz 2012), or stimuli that involve rule-driven input-function-output systems versus objects that allow for empathizing activities (Baron-Cohen 2003). Instead, they construct masculine, feminine, and gender-neutral categories on the basis of their assumed or observed popularity with girls and boys in a particular time and place (Bleier 1984). Then, if an assumed "boy toy" is unexpectedly popular with girls, it is removed from the stimulus set (Pasterski et al. 2005).

Brain organization theory can be understood as seeking to elucidate proximal mechanisms predicted by mainstream evolutionary theory: that is, how does an inherited male or female genome give rise to evolved sex-linked mating behavior? Accordingly, research conducted within the framework of brain organization theory has typically treated the environment as noise or a background condition for the development of sex differences in the brain. As McCarthy and Arnold (2011, p. 677) have noted: "Biological theories of sexual differentiation have largely under-emphasized or even excluded the differential effect of sex-specific environments." However, mainstream evolutionary theory has been criticized for neglecting additional, nongenetic channels of inheritance available to organisms—namely, epigenetic, behavioral, and symbolic processes (Jablonka and Lamb 2005). Although it remains a topic of controversy within evolutionary science, it has been argued that there has been insufficient consideration of the contribution of nongenetic forms of inheritance (e.g., ecological niche, food sources, parental care, uterine environment, gravity). In the case of our own species, newborns also reliably inherit a rich (and gendered) culture.

"Developmental systems theory" posits that all four channels of inheritance can contribute to the phenotypic variation on which natural selection acts, and to the stable cross-generation transfer of adaptive traits (Oyama et al. 2003). Indeed, the examples above have all been linked to such effects (for descriptions of these examples and others, see Griffiths 2001; Moore 2003; Fine et al. 2017). For instance, newborn rat pups dis-

Cite this article as *Cold Spring Harb Perspect Biol* doi: 10.1101/cshperspect.a039115

play the adaptive behavior of approaching and latching on to the mother's nipples shortly after birth. However, this early response is contingent on odors experienced in utero and postnatally. If a substance is injected into the amniotic fluid that gives it a lemony smell, pups exposed to that smell in utero, as well as briefly immediately after birth, will orient only to nipples with the same scent. Control rats, by contrast, orient only to a naturally scented nipple (Pedersen and Blass 1982; described in Moore 2003).

From a developmental systems theory framework, the individual's experiences, including sex-specific ones, enjoy the status of playing a potentially causal role in the development and cross-generational transfer of adaptive traits—an example of how framework can influence research questions. For example, working from a developmental systems theory framework, Moore (1984) generated data indicating indirect sex effects on brain and behavior via the differential treatment of male and female rat pups by dams in response to higher testosterone levels in male pups' urine. As Moore (1992, pp. 173–174) noted:

> These results are exciting because they are the first demonstration that predictable, species-typical, afferent input from a natural source can affect the adult morphology of a sexually dimorphic nucleus in the central nervous system. They demonstrate that stimulation provided by a dam during the normal course of caring for developing young can contribute to the differentiation of neural mechanisms that underlie masculine sexual behavior. Variation among the dams that provide this stimulation can produce individual differences in neural mechanisms among males that may, in turn, affect their reproductive success. Finally, the results lead one to conclude that reliable differences in the stimulation provided to males and females contribute to the sexual dissimilarity of nervous system morphology.

As Fine et al. (2017) have argued, applying a developmental systems theory framework to human sex differences challenges the implicit assumption that sex-linked adaptive human traits persist because they are passed on via inherited biological sex (as opposed to other channels of inheritance). In other words, it is possible that for some sex-linked behavioral traits "what is genetically selected for and inherited is the ability to quickly acquire or learn adaptive behaviors, whereas the actual content of the behavior thus acquired may depend on sex-differentiated environmental processes" (Fine et al. 2017, p. 671.) In line with this possibility, unlike control children, girls with CAH do not show a "gender-appropriate" preference for items in pairs of objects where one item has been explicitly or implicitly labeled as being "for girls," and the other "for boys" (e.g., a toy cow versus a toy horse) (Hines et al. 2016). Hines et al. (2016) note that this suggests the possibility that elevated prenatal androgen levels may contribute to the increased interest in male-typical activities observed in this population via some kind of indirect or developmental effect on self-socialization processes. However, it also remains an open possibility that the latter mechanism, in combination with a sex-differentiated environment, fully explains prior findings.

Fair Criticism versus Politicization

What is fair criticism and what is harmful politicization in conflicts over background assumptions, frameworks, and dominant methodologies? As the brain organization theory example shows, criticism of these elements of research can play a legitimate and valuable role. Indeed, historians and philosophers of science have placed considerable importance on such exchanges for achieving greater scientific objectivity (see especially Longino 1990). For example, in her analysis of this rich and complex construct, Douglas (2009, pp. 127–128) offers an account of so-called "interactive objectivity," whereby:

> Instead of immediately assenting to an observation account, the participants are required to argue with each other, to ferret out the sources of their disagreements. It is in the spirit of this sense that we require that scientific data be shared, theories discussed, models be open to examination, and, if possible, experiments replicated. The open community of discussion has long been considered crucial for science. The hope is that by keeping scientific discourse open to scrutiny, the most idiosyncratic biases and blinders can be eliminated.

An interesting implication is that politicized science in the area of sex differences in the brain can potentially manifest in the absence of alternative research programs that draw on different frameworks, background assumptions, methodologies, and/or a mainstream scientific community that fails to properly engage with such accounts (Eigi 2012; Okruhlik 2013). (Of course, the same point can be made in relation to social science research programs focused on social influences on sex differences in behavior.) As Oreskes (2019, p. 249) has argued: "Diversity is crucial because, *ceteris paribus,* it increases the odds that any particular claim has been examined from many angles and potential shortcomings revealed. Homogenous groups often fail to recognize their shared biases." For example, inadvertently, the standard method for selecting toys in studies of girls with CAH has protected from revision the underlying background assumption that such toys are universally and timelessly appealing to a masculinized brain. This is a situation that "[distorts] the very thing we value in science – that it can revise itself in the face of evidence that goes against dearly held theories" (Douglas 2009, p. 101).

However, conflicts of this kind do not always enhance objectivity. For example, it is certainly possible that, either intentionally or inadvertently, a scientist's or academic's political, economic (or other) motives may lead them to ignore or remain blind to disconfirming evidence, empirical inadequacies in their own background assumptions, or the merits in others'. In Douglas's terminology, this would count as a failure of "detached objectivity." Claims of failure of detached objectivity are sometimes leveled against feminist critics, but they need to be documented and diagnosed, rather than simply assumed (Fine 2020). This criterion requires sincere and charitable efforts to engage and identify points of agreement and disagreement. Moreover, it should be noted that inadequate openness to disconfirming evidence is a possibility for any scientist invested in a particular framework, method, or theory, which is to say many scientists. Symptoms of failure of detached objectivity include commentary that stubbornly misrepresents or mischaracterizes the views of opponents, and persistent unresponsiveness to counterarguments and counterevidence (see Oreskes 2019).

CONFLICT 3: INDUCTIVE RISK AND SOCIAL VALUES

Scientists must decide whether to accept a particular hypothesis on the basis of available evidence. Such decisions may lead to correct "true-negative" or "true-positive" outcomes, or to erroneous "false-positive" or "false-negative" outcomes. The inductive nature of science means that empirical claims and scientific theories are always provisional (although to a lesser degree when there is largely consensus within the scientific community). Thus, a third kind of conflict that can occur concerns how inductive risk should be handled: what degree of uncertainty is tolerable in accepting or rejecting a particular hypothesis?

A common form of feminist research argues that a hypothesis, with socially regressive implications, is accepted on the basis of poor quality, misinterpreted, or overinterpreted evidence (e.g., Fine 2010; Bluhm 2013). Thus, one possible complaint is that science is not meeting expected standards. A second possible position is that higher than normal standards of evidence are necessary for the acceptance of hypotheses that could have harmful consequences for already marginalized groups if wrongly accepted as true (e.g., Kitcher 1985; Fausto-Sterling 1992). However, since such harms will be perceived through the lens of prior assumptions, ideologies, and value systems, there may be little consensus regarding which erroneous claims would be most harmful (e.g., M Del Giudice, forthcoming), and thus require raised standards of evidence. Moreover, should those standards become unrealistically or impossibly high, this approach to managing inductive risk could become a de facto form of suppression of academic freedom. If no degree of uncertainty were tolerable to accept a particular hypothesis, then certain conclusions effectively become out of bounds. Such a situation would strongly discourage lines of inquiry designed to test "disallowed" hypotheses.

These are all valid concerns that raise difficult issues. For example, even if we agree that all research should be held to the same standard regardless of the conclusions drawn, it is not always clear what that standard should be (for discussion of the impossibility of a universal standard of evidence within science, see Douglas [2011]). Particularly for research using novel methods, there may be no clear disciplinary standard to use as a yardstick for acceptance of claims. In these situations, scientists, reviewers, editors, and postpublication commentators must make their own judgments as to whether the risk of error is tolerable. Another issue is that critics might be justified in arguing that typical disciplinary standards are too low. For example, both the 2D:4D digit ratio and amniotic testosterone have often been used as proxies for prenatal testosterone exposure, but evidence points to them being too insensitive for use as an individual difference measure (Herbert 2015; Spencer et al. 2021). Or consider that standard practices in reporting of sex differences in neuroimaging studies appear to have given rise to a proliferation of false-positive errors (Fine and Fidler 2015; David et al. 2018).

Nor is it clear that a single disciplinary standard can be applied to all studies, in the absence of any further judgment on the part of the scientist. Returning to the *Journal of Neuroscience Research* policy mentioned earlier, this guideline strongly encourages authors to consider sex as a biological variable, even in cases where studies are not designed to examine sex differences, and/or there is no existing literature indicating sex influences in the domain of interest (Prager 2017). Prager (2017, p. 95) argues that "the real risk of false-positive errors … is balanced by the equal or greater risk of false-negative errors from failing to consider possible sex influences." However, the statistical risks of false-positive versus false-negative errors depend on factors that will vary across studies, such as statistical power (which will be reduced by the addition of sex as a variable), variability, effect size, and the likelihood that the hypothesis is actually true (Héroux 2017). Thus, the journal policy will, in practice, give rise to a highly inconsistent standard in terms of the uncertainty of the claim. This is also in tension with recent recommendations for increasing the reproducibility of scientific research, such as that "researchers should transparently report and justify all choices they make when designing a study, including the alpha level" (Lakens et al. 2018,

p. 168). Moreover, scientists working with different frameworks and background assumptions may approach data with different priors (e.g., the likely existence or magnitude of a sex difference) that will also affect scientists' individual assessments of uncertainty.

The situation becomes even more complex when one introduces the idea that scientists need greater certainty to make claims for which there are clear and significant consequences of error (for brief historical overview of the development of this position in philosophy of science, see Douglas 2009). To be clear, the argument is not that such concerns determine which claims are "allowable"—a so-called "direct role," as Douglas (2009, p. 113) has argued:

> It is one thing to argue that a view is not sufficiently supported by the available evidence. It is quite another to suggest that the evidence be ignored because one does not like its implications. In the former case, we can argue about the quality of the evidence, how much should be enough, and where the burdens of proof should lie. These are debates we should be having about contentious, technically based issues. In the latter case, we can only stand aghast at the deliberate attempt to wish the world away. Allowing a direct role for values is a common way to politicize science, and to undermine the reason we value it at all: to provide us with reliable knowledge about the world.

Even if we agree in principle with the possibility of shifting burdens of evidence, depending on the consequences of error, this is no simple task. The complex and value-laden nature of risk perception (e.g., Kunreuther and Slovic 1996) suggests that, in practice, it may be difficult to disentangle perceptions of uncertainty from perceptions of the harms likely to ensue from different kinds of error (Elliott 2011). Nonetheless, greater transparency about values might facilitate interactive objectivity, by reducing the extent to which disagreements about values are conflated with disagreements over data (Douglas 2008).

Fair Criticism versus Politicization

Given all of the above, it is likely that many conflicts over the neuroscience of sex differences concern the handling of inductive risk. What, then, counts as fair criticism, and what is politicization? Inevitably, some cases will be (or at least should be) clearer than others. Decisively falling into the category of fair criticism is research that critiques or rejects theories or claims that are internally inconsistent (i.e., predicts both X and not X), or that make predictions that are clearly inconsistent with empirical evidence. As Douglas (2009, p. 94) notes, any theory or hypothesis that does not meet these criteria fails to meet acceptable standards of science. Moreover, there should also be relatively little concern over criticism or rejection of research that, say, deploys methods or equipment widely agreed to be unreliable, or that uses clearly improper statistical techniques. Beyond these more clear-cut cases, however, will be those for which there can be reasonable disagreement. What, to some, looks like the responsible handling of uncertainty will, to others, appear as the suppression or rejection of politically inconvenient findings. Conversely, what looks to some like the publication of interesting research will, to others, look like poor quality science enjoying undue credence due to its resonance with prevailing gender stereotypes or ideology.

However, at a certain point, rejection of a claim on the grounds of the possibility of error moves into the territory of politicization, albeit that there is no bright line marking the boundary. Indeed, it may take time to reveal whether social values (in either direction) are playing an inappropriate role in decisions about the handling of uncertainty. For example, if new, improved, or convergent evidence (and subsequent reduction of uncertainty) does not shift a scientist's position, this may indicate that social values are indeed standing in for evidence (Douglas 2015; Oreskes 2019). However, establishing whether this is the case requires sustained scientific attention on a particular empirical claim that genuinely leads to progressive reductions of uncertainty. For various reasons, this may not always be the case since, for example, scientists often enjoy greater rewards (and funding) for producing novel findings than replications, or alternatively may find it difficult to publish findings that challenge dominant theories or methodologies (see Gleditsch 2021). In other words, a continuing impasse over whether a particular hypothesis has adequate warrant may reflect continuing uncertainty, rather than its politicization.

CONCLUDING REMARKS

I have summarized three kinds of conflict that can arise in the neuroscience of sex differences in the human brain: academic freedom versus gender equality; frameworks, background assumptions, and dominant methodologies; and the handling of inductive risk (Fig. 1). While distinct in principle, in practice they will often overlap. For example, different background assumptions may give rise to different assessments of the degree of uncertainty of an empirical claim. This, in turn, may result in a different conclusion about handling inductive risk. Or concerns about significant inductive risks, tilted toward harmful error, could potentially give rise to arguments in favor of constraining academic freedom in the interests of gender equality. Nonetheless, these are conceptually distinct ways in which interlocutors can disagree, and conflicts are more likely to be productive when there is greater clarity on the source of disagreement.

Throughout this review, I have also tried to clarify the boundaries between fair criticism and politicization, although there may not always be a bright line between the two. With regard to conflicts over academic freedom versus equality, it is of course perfectly legitimate for scientists and other scholars to criticize the presence or absence of restrictions on academic freedom, or strong institutional shaping of the direction of research programs, in the interests of equality. What is certainly not legitimate are attempts to impose such constraints de facto through scholarship suppression. Whereas freedom of expression allows anyone, including academics, to call for scholarship suppression, protection of academic freedom requires that conference organizers, journal editors, and university administrators do not yield to such demands (Redstone and Villasenor 2020). Academic freedom is compatible with concepts of social equality and inclusion within universities (Anderson 1995), and universities and scientific institu-

 Cite this article as *Cold Spring Harb Perspect Biol* doi: 10.1101/cshperspect.a039115

tions can maintain their commitment to both by ensuring free and open academic debate that works to actively include a wide range of perspectives and voices. Indeed, Longino (2002) has argued that the best way to achieve objectivity, including in politically sensitive domains, is to ensure that the scientific community nurtures, and is open to, multiple perspectives, and maintains academic venues that support the exchange and uptake of criticism. Cogent arguments have been made that this approach, rather than suppression of academic freedom, would best address concerns about negative social consequences of biased research (Eigi 2012).

For conflicts over background assumptions, identifying where fair criticism shades into politicization requires care, humility, openness to opposing points of view, and a charitable mindset. These virtues are probably more likely to be displayed to the extent that interlocutors are confident that they are, or will be, reciprocated by those with opposing views. It could also be helpful, to this end, for scientists to abide by professional norms of conduct—in particular, the avoidance of ad hominem commentary—when they engage in scientific debates in nonprofessional fora such as social media.

The boundary between fair criticism and politicization is perhaps the most blurred when it comes to the handling of inductive risk. Here, disagreements may concern the degree of uncertainty of an empirical claim, the appropriate burden of proof, and/or the social consequences of error. Disentangling these different possible contributions to a specific disagreement is a difficult, perhaps impossible, task at the individual level. Douglas (2009, p. 123) has proposed that scientists attempt to cultivate "value-neutral objectivity," whereby "scientists take a position that is balanced or neutral with respect to a spectrum of values." However, she argues that this is not always the appropriate strategy: for example, "if racist or sexist values are at one end of the value continuum, value-neutrality would not be a good idea" (p. 124). Whereas this seems reasonable with respect to male supremacist values, beyond this there may not be agreement among scientists as to what values are rightly categorized as sexist (e.g., M Del Giudice, forthcoming).

In combination with the fact that peer review is an imperfect process, there will inevitably be controversial cases of the publication of poor quality, overinterpreted, or misinterpreted articles, perhaps with regressive social implications. Whereas each case has to be considered on its merits, retractions in response to external pressure (such as petitions)—in the absence of data fraud, other forms of academic misconduct, or clear empirical error—give genuine cause for concern. These events undermine trust that scientific institutions will not capitulate to political or economic interests. Moreover, in cases that are borderline (and even those that are not), the optics to many will be that some scientists (or academic institutions) are indeed politicizing science. This perception is distinctly unhelpful for those concerned with improving the rigor of the science of group differences, as it makes it easier and more likely for even valid critiques to be dismissed as politicization. At the same time, it is useful to recognize that the reality of academic publishing practices often fall short of ideals of free and open debate, and that greater efforts to enhance these opportunities, and better align incentives to promote them, would be beneficial (see Gleditsch 2021).

Conflicts over the neuroscience of sex differences are unlikely to abate any time soon—and this is how it should be. All scholars, as well as those who manage and lead research institutions, have a role to play in ensuring that research in this area remains fairly criticized, without being politicized.

ACKNOWLEDGMENTS

I am grateful to Marc Breedlove, Russell Blackford, Fiona Fidler, Daphna Joel, and two anonymous reviewers for their very helpful comments on earlier versions of the manuscript.

REFERENCES

Alexander GM. 2003. An evolutionary perspective of sex-typed toy preferences: pink, blue, and the brain. *Arch Sex Behav* **32:** 7–14. doi:10.1023/A:1021833110722

Alexander GM, Saenz J. 2012. Early androgens, activity levels and toy choices of children in the second year of life.

Horm Behav **62**: 500–504. doi:10.1016/j.yhbeh.2012.08.008

Anderson ES. 1995. The democratic university: the role of justice in the production of knowledge. *Soc Philos Policy* **12**: 186–219. doi:10.1017/S0265052500004726

Bailey JM. 2019. How to ruin sex research. *Arch Sex Behav* **48**: 1007–1011. doi:10.1007/s10508-019-1420-y

Bainbridge R, Tsey K, McCalman J, Kinchin I, Saunders V, Watkin Lui F, Cadet-James Y, Miller A, Lawson K. 2015. No one's discussing the elephant in the room: contemplating questions of research impact and benefit in Aboriginal and Torres Strait Islander Australian health research. *BMC Public Health* **15**: 696. doi:10.1186/s12889-015-2052-3

Barinaga M. 2000. Asilomar revisited: lessons for today? *Science* **287**: 1584–1585. doi:10.1126/science.287.5458.1584

Baron-Cohen S. 2003. *The essential difference: the truth about the male and female brain*. Basic Books, New York.

Baron-Cohen S. 2010. Book review: Delusions of gender – "neurosexism," biology and politics. *The Psychologist* **23**: 904–905.

Bleier R. 1984. *Science and gender: a critique of biology and its theories on women*. Pergamon, Oxford.

Bluhm R. 2013. Self-fulfilling prophecies: the influence of gender stereotypes on functional neuroimaging research on emotion. *Hypatia* **28**: 870–886. doi:10.1111/j.1527-2001.2012.01311.x

Bluhm R. 2020. Neurosexism and our understanding of sex differences in the brain. In *The Routledge handbook of feminist philosophy of science*, 1st ed. (ed. Crasnow S, Intemann K), pp. 316–327. Routledge, New York.

Bluhm R, Jacobson AJ, Maibom HL. 2012. *Neurofeminism: issues at the intersection of feminist theory and cognitive science*. Macmillan, New York.

Cahill L. 2014. Equal ≠ the same: sex differences in the human brain. Dana Foundation, https://dana.org/article/equal-≠-the-same-sex-differences-in-the-human-brain

Cahill L. 2017. An issue whose time has come. *J Neurosci Res* **95**: 12–13. doi:10.1002/jnr.23972

Cummins E. 2021. UCLA pauses "unethical" study designed to mentally distress trans people. *Vice*, https://www.vice.com/en/article/m7a4n4/ucla-pauses-unethical-study-designed-to-mentally-distress-trans-people

David SP, Naudet F, Laude J, Radua J, Fusar-Poli P, Chu I, Stefanick ML, Ioannidis JPA. 2018. Potential reporting bias in neuroimaging studies of sex differences. *Sci Rep* **8**: 6082. doi:10.1038/s41598-018-23976-1

Del Giudice M (forthcoming). Ideological bias in the psychology of sex and gender. In *Political bias in psychology: nature, scope, and solutions* (ed. Frisby CL, et al.). Springer, New York.

Doell RG, Longino HE. 1988. Sex hormones and human behavior: a critique of the linear model. *J Homosex* **15**: 55–78. doi:10.1300/J082v15n03_03

Douglas H. 2008. The role of values in expert reasoning. *Public Aff Q* **22**: 1–18.

Douglas HE. 2009. *Science, policy, and the value-free ideal*. University of Pittsburgh Press, Pittsburgh, PA.

Douglas H. 2011. Bullshit at the interface of science and policy: global warming, toxic substances, and other pesky problems. In *Bullshit and philosophy: guaranteed to get perfect results every time* (ed. Hardcastle GL, Reisch GA). Open Court, Chicago.

Douglas H. 2015. Politics and science: untangling values, ideologies, and reasons. *Ann Am Acad Pol Soc Sci* **658**: 296–306. doi:10.1177/0002716214557237

Eigi J. 2012. Two Millian arguments: using Helen Longino's approach to solve the problems Philip Kitcher targeted with his argument on freedom of inquiry. *Studia Philosophica Estonica* **5**: 44–63. doi:10.12697/spe.2012.5.1.03

Eliot L. 2011. The trouble with sex differences. *Neuron* **72**: 895–898. doi:10.1016/j.neuron.2011.12.001

Eliot L, Richardson SS. 2016. Sex in context: limitations of animal studies for addressing human sex/gender neurobehavioral health disparities. *J Neurosci* **36**: 11823–11830. doi:10.1523/JNEUROSCI.1391-16.2016

Elliott KC. 2011. Direct and indirect roles for values in science. *Philos Sci* **78**: 303–324. doi:10.1086/659222

Fausto-Sterling A. 1992. *Myths of gender: biological theories about women and men*, 2nd ed. BasicBooks, New York.

Fazackerley A. 2018. UK universities struggle to deal with "toxic" trans rights row. *The Guardian*, http://www.theguardian.com/education/2018/oct/30/uk-universities-struggle-to-deal-with-toxic-trans-rights-row

Fine C. 2010. *Delusions of gender: how our minds, society, and neurosexism create difference*, 1st ed. W. W. Norton, New York.

Fine C. 2013. Is there neurosexism in functional neuroimaging investigations of sex differences? *Neuroethics* **6**: 369–409. doi:10.1007/s12152-012-9169-1

Fine C. 2015. Neuroscience, gender, and "development to" and "from": the example of toy preferences. In *Handbook of neuroethics* (ed. Clausen J, Levy N), pp. 1737–1755. Springer, Dordercht, Netherlands.

Fine C. 2020. Constructing unnecessary barriers to constructive scientific debate: a response to Buss and von Hippel (2018). *Arch Sci Psychol* **8**: 5–10.

Fine C, Fidler F. 2015. Sex and power: why sex/gender neuroscience should motivate statistical reform. In *Handbook of neuroethics* (ed. Clausen J, Levy N), pp. 1447–1462. Springer, Dordercht, Netherlands.

Fine C, Dupré J, Joel D. 2017. Sex-linked behavior: evolution, stability, and variability. *Trends Cogn Sci* **21**: 666–673. doi:10.1016/j.tics.2017.06.012

Gleditsch KS. 2021. Houston, we have a problem: enhancing academic freedom and transparency in publishing through post-publication debate. *Political Stud Rev* **19**: 428–434. doi:10.1177/1478929919889309

Greenberg DM, Warrier V, Allison C, Baron-Cohen S. 2018. Testing the empathizing–systemizing theory of sex differences and the extreme male brain theory of autism in half a million people. *Proc Natl Acad Sci* **115**: 12152–12157. doi:10.1073/pnas.1811032115

Griffiths PE. 2001. What is innateness? *Monist* **85**: 70–85. doi:10.5840/monist20028518

Grossi G, Fine C. 2012. The role of fetal testosterone in the development of the "essential difference" between the sexes: some essential issues. In *Neurofeminism* (ed.

Bluhm R, et al.), pp. 73–104. Palgrave Macmillan, London.

Herbert J. 2015. *Testosterone: sex, power, and the will to win*, 1st ed. Oxford University Press, Oxford.

Héroux M. 2017. Why most published findings are false: revisiting the Ioannidis argument. *Scientifically Sound*, https://scientificallysound.org/2017/10/04/most-published-findings-are-false

Hines M. 2015. Gendered development. In *Handbook of child psychology and developmental science* (ed. Lerner RM), pp. 1–46. Wiley, Hoboken, NJ.

Hines M, Pasterski V, Spencer D, Neufeld S, Patalay P, Hindmarsh PC, Hughes IA, Acerini CL. 2016. Prenatal androgen exposure alters girls' responses to information indicating gender-appropriate behaviour. *Philos Trans R Soc Lond B Biol Sci* **371**: 20150125. doi:10.1098/rstb.2015.0125

Jablonka E, Lamb MJ. 2005. *Evolution in four dimensions: genetic, epigenetic, behavioral, and symbolic variation in the history of life*. MIT Press, Cambridge, MA.

Joel D, Fausto-Sterling A. 2016. Beyond sex differences: new approaches for thinking about variation in brain structure and function. *Philos Trans R Soc Lond B Biol Sci* **371**: 20150451. doi:10.1098/rstb.2015.0451

Joel D, Kaiser A, Richardson SS, Ritz SA, Roy D, Subramaniam B. 2015. A discussion on experiments and experimentation: NIH to balance sex in cell and animal studies. *Catalyst: Feminism, Theory, Technoscience* **1**: 1–13. doi:10.28968/cftt.v1i1.28821

Johnson DG. 1996. Forbidden knowledge and science as professional activity. *Monist* **79**: 197–217. doi:10.5840/monist19967927

Jordan-Young RM. 2011. *Brain storm: the flaws in the science of sex differences*. Harvard University Press, Cambridge, MA.

Jordan-Young RM. 2012. Hormones, context, and "brain gender": a review of evidence from congenital adrenal hyperplasia. *Soc Sci Med* **74**: 1738–1744. doi:10.1016/j.socscimed.2011.08.026

Kitcher P. 1985. *Vaulting ambition: sociobiology and the quest for human nature*. MIT Press, Cambridge, MA.

Kitcher P. 2001. *Science, truth, and democracy*. Oxford University Press, Oxford.

Kourany JA. 2016. Should some knowledge be forbidden? The case of cognitive differences research. *Philos Sci* **83**: 779–790. doi:10.1086/687863

Kunreuther H, Slovic P. 1996. Science, values, and risk. *Ann Am Acad Pol Soc Sci* **545**: 116–125. doi:10.1177/0002716296545001012

Lakens D, Adolfi FG, Albers CJ, Anvari F, Apps MAJ, Argamon SE, Baguley T, Becker RB, Benning SD, Bradford DE, et al. 2018. Justify your alpha. *Nat Hum Behav* **2**: 168–171. doi:10.1038/s41562-018-0311-x

Lilienfeld SO. 2002. When worlds collide: social science, politics, and the Rind et al. (1998). Child sexual abuse meta-analysis. *Am Psychol* **57**: 176–188. doi:10.1037/0003-066X.57.3.176

Lockhart JW. 2020. "A large and longstanding body": historical authority in the science of sex. In *Far-right revisionism and the end of history* (ed. Valencia-García LD). Routledge, New York.

Longino HE. 1990. *Science as social knowledge: values and objectivity in scientific inquiry*. Princeton University Press, Princeton, NJ.

Longino HE. 2002. *The fate of knowledge*. Princeton University Press, Princeton, NJ.

Longino H, Doell R. 1983. Body, bias, and behavior: a comparative analysis of reasoning in two areas of biological science. *Signs (Chic)* **9**: 206–227. doi:10.1086/494044

McCarthy MM, Arnold AP. 2011. Reframing sexual differentiation of the brain. *Nat Neurosci* **14**: 677–683. doi:10.1038/nn.2834

Mill JS. 1859. *On liberty*. John W. Parker and Son, London (reprinted 1991, ed. Gray J. Oxford University Press, Oxford).

Moore CL. 1984. Maternal contributions to the development of masculine sexual behavior in laboratory rats. *Dev Psychobiol* **17**: 347–356. doi:10.1002/dev.420170403

Moore CL. 1992. The role of maternal stimulation in the development of sexual behavior and its neural basis. *Ann NY Acad Sci* **662**: 160–177. doi:10.1111/j.1749-6632.1992.tb22859.x

Moore CL. 2003. Evolution, development, and the individual acquisition of traits: what we've learned since Baldwin. In *Evolution and learning: the Baldwin effect reconsidered*, pp. 115–139. MIT Press, Cambridge, MA.

National Health and Medical Research Council (Australia). 2018a. Ethical conduct in research with Aboriginal and Torres Strait Islander peoples and communities: Guidelines for researchers and stakeholders. NHMRC, Canberra, Australia.

National Health and Medical Research Council (Australia). 2018b. Keeping research on track II: a companion document to ethical conduct in research with Aboriginal and Torres Strait Islander peoples and communities: guidelines for researchers and stakeholders. NHMRC, Canberra, Australia.

National Research Council. 2004. *Biotechnology research in an age of terrorism*. National Academies, Washington, DC.

NIH. 2015. Consideration of sex as a biological variable in NIH-funded research. National Institutes of Health, https://grants.nih.gov/grants/guide/notice-files/NOT-OD-15-102.html

Okruhlik K. 2013. Gender and the biological sciences. *Can J Philos*, https://www.tandfonline.com/doi/abs/10.1080/00455091.1994.10717393

Oreskes N. 2019. *Why trust science?* Princeton University Press, Princeton, NJ.

Oyama S, Griffiths PE, Gray RD. 2003. *Cycles of contingency: developmental systems and evolution*. MIT Press, Cambridge, MA.

Pasterski VL, Geffner ME, Brain C, Hindmarsh P, Brook C, Hines M. 2005. Prenatal hormones and postnatal socialization by parents as determinants of male-typical toy play in girls with congenital adrenal hyperplasia. *Child Dev* **76**: 264–278. doi:10.1111/j.1467-8624.2005.00843.x

Pedersen PE, Blass EM. 1982. Prenatal and postnatal determinants of the 1st suckling episode in albino rats. *Dev Psychobiol* **15**: 349–355. doi:10.1002/dev.420150407

Prager EM. 2017. Addressing sex as a biological variable: editor's column. *J Neurosci Res* **95:** 11–11. doi:10.1002/jnr.23979

Redstone I, Villasenor J. 2020. *Unassailable ideas: how unwritten rules and social media shape discourse in American higher education.* Oxford University Press, Oxford.

Richardson SS. 2010. Feminist philosophy of science: history, contributions, and challenges. *Synthese* **177:** 337–362. doi:10.1007/s11229-010-9791-6

Richardson SS, Reiches M, Shattuck-Heidorn H, LaBonte ML, Consoli T. 2015. Opinion: focus on preclinical sex differences will not address women's and men's health disparities. *Proc Natl Acad Sci* **112:** 13419–13420. doi:10.1073/pnas.1516958112

Rippon G, Jordan Young R, Kaiser A, Fine C. 2014. Recommendations for sex/gender neuroimaging research: key principles and implications for research design, analysis, and interpretation. *Front Hum Neurosci* **8:** 650. doi:10.3389/fnhum.2014.00650

Rosenstock L, Lee LJ. 2002. Attacks on science: the risks to evidence-based policy. *Am J Public Health (NY)* **92:** 14–18. doi:10.2105/AJPH.92.1.14

Russett CE. 1989. *Sexual science: the Victorian construction of womanhood.* Harvard University Press, Cambridge, MA.

Schiebinger L. 1996. *The mind has no sex? Women in the origins of modern science.* Harvard University Press, Cambridge, MA.

Singal J. 2020. Can science ever debate trans issues? *UnHerd,* https://unherd.com/2020/06/eneuro

Spencer D, Pasterski V, Neufeld SAS, Glover V, O'Connor TG, Hindmarsh PC, Hughes IA, Acerini CL, Hines M. 2021. Prenatal androgen exposure and children's gender-typed behavior and toy and playmate preferences. *Horm Behav* **127:** 104889. doi:10.1016/j.yhbeh.2020.104889

Stevens ST, Jussim L, Honeycutt N. 2020. Scholarship suppression: theoretical perspectives and emerging trends. *Societies* **10:** 82. doi:10.3390/soc10040082

Stone A, Evans C. 2021. *Open minds: academic freedom and freedom of speech of Australia.* La Trobe University Press, Carlton, Australia.

Suissa J, Sullivan A. 2021. The gender wars, academic freedom and education. *J Philos Educ* **55:** 55–82.

Tavris C. 1993. *The mismeasure of woman.* Touchstone, New York.

van Anders SM. 2013. Beyond masculinity: testosterone, gender/sex, and human social behavior in a comparative context. *Front Neuroendocrinol* **34:** 198–210. doi:10.1016/j.yfrne.2013.07.001

Whipple T. 2016. Sexism fears hamper brain research. *The Times* (November 29).

White PC, Speiser PW. 2000. Congenital adrenal hyperplasia due to 21-hydroxylase deficiency. *Endocr Rev* **21:** 245–291.

Pervasive Neglect of Sex Differences in Biomedical Research

Irving Zucker,[1,2] Brian J. Prendergast,[3] and Annaliese K. Beery[2,4]

[1]Department of Psychology, University of California, Berkeley, Berkeley, California 94720, USA

[2]Department of Integrative Biology, University of California, Berkeley, Berkeley, California 94720, USA

[3]Department of Psychology and Committee on Neurobiology, University of Chicago, Chicago, Illinois 60637, USA

[4]Program in Neuroscience, Departments of Psychology and Biology, Smith College, Northampton, Massachusetts 01063, USA

Correspondence: irvzuck@berkeley.edu

Females have long been underrepresented in preclinical research and clinical drug trials. Directives by the U.S. National Institutes of Health have increased female participation in research protocols, although analysis of outcomes by sex remains infrequent. The long-held view that traits of female rats and mice are more variable than those of males is discredited, supporting equal representation of both sexes in most studies. Drug pharmacokinetic analysis reveals that, among subjects administered a standard drug dose, women are exposed to higher blood drug concentrations and longer drug elimination times. This contributes to increased adverse drug reactions in women and suggests that women are routinely overmedicated and should be administered lower drug doses than men. The past decade has seen progress in female inclusion, but key subsequent steps such as sex-based analysis and sex-specific drug dosing remain to be implemented.

In the not too distant past, men were considered representative of the human species; differences from the male norm were viewed as atypical or abnormal, just one aspect of a broader sexism that ranks among the most pervasive human prejudices (Perry and Albee 1998). The notion that women and girls were inferior to men and boys was in play when agriculture and sedentary cultures emerged (Ananthaswamy and Douglas 2018), with the status of women lower than that of men from the dawn of recorded history (Rosen 1971; Morsink 1979). Sex bias persists today in virtually all walks of life, creating an environment that disadvantages women.

Women and nonhuman female mammals have been given short shrift in biomedical research. Until recently, the research community labored under the misguided assumption that information garnered from studies of males could be generalized without modification to females. Since then, sex differences in mechanisms underlying basic biological processes, from pain signaling (e.g., Mogil et al. 2003; Sorge et al. 2015) to synaptic inhibition (Huang and Woolley 2012), to drug metabolism (described below), have been detailed with important consequences for women's health (Heinrich 2001; Klein et al. 2015).

The belief that hormonal variations associated with estrous and menstrual cycles renders females more variable than males—now discredited (Mogil and Chanda 2005; Prendergast et al. 2014; Becker et al. 2016)—discouraged inclusion of women and female rodents in experimental protocols, negatively impacting the quality of medical care for women. Mazure and Jones (2015) provide a detailed chronology of changes instituted by the U.S. National Institutes of Health (NIH) to address sex bias in human medical research, and provide a comprehensive list of actions needed to remove remaining barriers to ensure appropriate consideration of sex as a biological variable (SABV) in biomedical studies.

Here we review historical and current trends of female inclusion in preclinical and clinical research, evolving policies aimed at increasing participation of females in biomedical research, and ongoing deficits in analysis with sex or gender as a factor. The assumption that periodic fluctuations in hormone secretion compromise female participation in scientific studies is discredited, and the pharmacological basis of human sex differences in adverse drug reactions (ADRs) is explored.

HISTORICAL ANALYSIS

Inclusion of both sexes in research studies has been consistently low (~15%) over a 100-year interval (Fig. 1A; Beery and Zucker 2011). In the early to mid-twentieth century, the majority (60%–80%) of articles dealing with nonhuman mammals failed to report the sex of research subjects used in biological research. Since then, the percent of articles omitting subject sex has markedly decreased, but has been accompanied by a concomitant increase in the percent of articles reporting the exclusive use of males.

Inclusion of women in clinically relevant research has been marginally to substantially better over the past half century (Beery and Zucker 2011). Between 1949 and 1989, 32%–45% of studies incorporated both male and female sub-

jects, with a marked increase to >60% in 1999 and 2009. Articles with sex unspecified declined from over 20% between 1949 and 1979, to ~7% between 1989 and 2009.

CHANGES IN THE PAST DECADE: SEX BIAS

In 2009, we documented extensive male bias in research on humans and nonhuman mammals in eight of ten surveyed biological subdisciplines (Beery and Zucker 2011). The ratio of articles reporting on only males versus only females was most skewed in the fields of neuroscience (5.5:1) and pharmacology (5:1)—two research domains with strong preclinical relevance. A female skew was present only in studies of reproduction (1:1.6), and in immunology (1:2.2). Subject sex was often not reported in publications in 2009. Sex was omitted in 22%–42% of articles in neuroscience and physiology; at least 92% of articles in the behavior, endocrinology, and pharmacology categories specified sex of experimental animals or tissues (Beery and Zucker 2011).

In 2016, a U.S. National Institutes of Health policy change (NOT-OD-15-102; NIH) required investigators to consider SABV in grant applications. Woitowich and Woodruff (2019) assessed the short-term impact of this directive by surveying attitudes about the 2016 policy and perceptions regarding its implementation among NIH study section members in 2016 and 2017. A majority of respondents considered it important for NIH-funded research to consider SABV and thought it would improve rigor and reproducibility of findings. The percentage of grant applications that successfully addressed and incorporated the policy increased over this span.

A follow-up study analyzed articles in nine biological disciplines in 2019 to compare with a similar survey a decade earlier, and to assess the extent of incorporation of SABV in the years after the 2016 NIH directive (Woitowich et al. 2020). The percent of studies that included both sexes increased across pooled subdisciplines (Fig. 1B), with significant gains in many but not all specific fields (e.g., neuroscience but not pharmacology). Overall, sex-inclusive research practices have increased since 2009, although male bias remains in

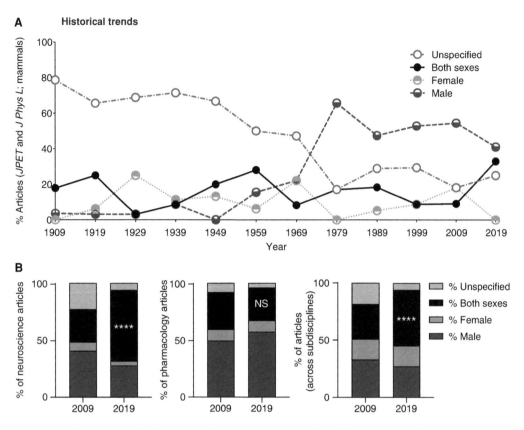

Figure 1. Sex bias in research subjects tested over time. (*A*) Sampling of mammalian studies published in two physiology journals over the past century reveals that including both sexes is not the norm. Decreased omission of the sexes of research subjects ("unspecified") over time coincides with an increase in reports of male-only studies. Historical data through 2009 are from Beery and Zucker (2011) (reprinted with permission from the authors). 2019 data from the *Journal of Physiology* (London) were obtained from the supplementary data file in Woitowich et al. (2020) (reprinted with permission from the authors) (note the *Journal of Pharmacology and Experimental Therapeutics* was not sampled for this year). (*B*) Across biological science subdisciplines surveyed, more 2009 studies examined only males than examined females or both sexes. By 2019, studies involving both sexes in at least one part of the study were significantly more common. This pattern was evident in some subdisciplines and not others. For example, there was an increase in sex-inclusive articles in neuroscience, but no such increase in pharmacology. ****$p < 0.0001$; NS = not significant.

several biological disciplines (Woitowich et al. 2020), and has increased in some fields (e.g., cardiovascular research [Ramirez et al. 2017]). In neuroscience, these gaps appear more prevalent for research performed in some species (e.g., rats, ferrets), but do not currently directly reflect funding source (Mamlouk et al. 2020). Most reports of single-sex male studies did not provide a rationale for excluding females (Woitowich et al. 2020), but justification for female exclusion still sometimes invoked the belief that females exhibit cyclicity-driven variability (now discredited, see below).

LACK OF CHANGE IN THE PAST DECADE: SEX-BASED ANALYSIS

Whereas inclusion of females has improved over the past decade, there has been no concomitant increase in the percentage of studies on both sexes that analyze results with sex or gender as a factor in analyses, and fewer than half of sex-inclusive studies make reference to any such analysis (Fig. 2; Mamlouk et al. 2020; Woitowich et al. 2020). Thus, improved female inclusion has not yet translated into increased reporting of female biology.

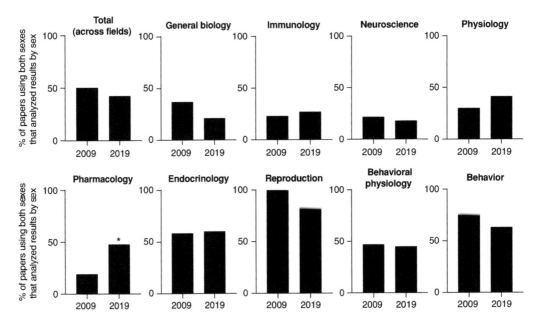

Figure 2. Lack of progress in analysis of findings by sex. While inclusion of both sexes increased from 2009 to 2019, there was no change in sex-specific analysis of the data: across fields, less than half of the studies that used both sexes reported any analysis by sex. Separate analysis of nine biological subdisciplines showed only the field of pharmacology increased in percentage of sex-based analyses, but this needs to be understood in the context of a field that has not increased inclusion of females. In other words, when that minority of papers in this field do include females (see Fig. 1B), they are likely to look for sex differences. (Data are from Beery and Zucker 2011 and Woitowich et al. 2020; reprinted, with permission, from the respective authors.)

The lack of sex-based analyses in sex-inclusive studies is potentially problematic. Males and females should not be pooled in analysis without—at a minimum—screening for sex differences (and reporting this screen); otherwise, experimental power may be lost (Beery 2018; Buch et al. 2019). When sex differences are present, inclusion of sex as a factor in analysis allows authors to identify these differences, as well as to increase the power to detect main effects of manipulations or treatments in the presence of sex differences in mean values (Beery 2018). Exploratory analysis of sex effects using factorial designs can assess the main effects of treatment and subject sex with effectively the same power as pairwise tests, without increased sample size (Collins et al. 2014; Buch et al. 2019). In the case where there is an interaction between sex and the main variable of interest, additional samples may be needed, and testing designed to capture sex differences will be biologically meaningful (Becker et al. 2005).

Fears of the need for doubled sample sizes (or more) in the presence of sex differences are thus unfounded. When sex differences are absent or when they are present without a sex × treatment interaction, no sample size increase is needed (Beery 2018; Buch et al. 2019). When both sex differences and a sex × treatment interaction are present (i.e., the situation requiring the largest sample size increases), a recent model revealed necessary increases of only 14%–33% under different conditions to include both sexes, even after statistical correction for the use of multiple factors (Buch et al. 2019).

The actual prevalence of sex differences in biology remains unknown, in part because there have been so few systematic surveys of the topic. This gap in reporting suggests that known sex differences are just the tip of the iceberg. To improve the validity of preclinical and clinical research, it is not enough simply to include females in research studies; one must also examine whether there are effects of sex/gender. Several guides for researchers to analyze sex differences now exist (Becker et al. 2005; Beltz et al. 2019; Buch et al. 2019).

FEMALE RODENTS ARE NOT A LIABILITY IN PRECLINICAL RESEARCH

The long-standing assumption that the estrous cycle renders females intrinsically more variable than males may be the single greatest contributor to the sex biases cataloged in the preceding section (Mogil and Chanda 2005). The underrepresentation of female animals in biomedical research is based on the assumption that, for any given trait, females exhibit more variability than males, and therefore must be tested at each of four stages of the estrous cycle to generate reliable data (Wald and Wu 2010). Satisfying this requirement would quadruple the number of females compared to studies employing males, thereby increasing experimental effort and cost. However, the question of whether such sex differences in variability even exist had not been addressed until recently.

In an assessment of more than 8000 individual measurements collected on 40 different mouse strains in three different laboratories, researchers found that females tested at random points in their estrous cycles were no more variable than males on an acute thermal pain test; no sex differences in variability measures were observed in acute and tonic chemical nociception tests (Mogil and Chanda 2005). This inference transcended measures of pain, evidenced by assessment of variability between male and female mice on ~10,000 morphological, physiological, and behavioral traits (Prendergast et al. 2014). This analysis focused on measures obtained without regard for the stage of the estrous cycle, thus maximizing any supposed female variability. Across this diverse array of traits, female variability was shown to be no greater than that of males even when estrous cycles were not monitored, thereby eliminating the barrier that fostered underrepresentation of female rodents in biomedical research (Fig. 3A).

The analysis also examined coefficients of variation (CVs) sorted into 30 broad trait categories but found no systematic pattern of sex-biased CVs—a similar conclusion was reached from an analysis of a set of >2 million data points, across 218 traits measured from >26,000 mice (International Mouse Phenotyping Consortium; Zajitschek et al. 2020). An evaluation of variability in gene expression, performed on 293 microarray data sets from mice and humans, found gene expression in males to be slightly more variable than that of females, although the difference was small (Fig. 3B; Itoh and Arnold 2015). The most common tissue source for this analysis was brain, which showed either a slight male bias in CV or no difference. Distinct patterns of sex differences in variance were identified in individual organs: in most cases CVs were higher in males (kidney, adrenal, skeletal muscle), but variance was greater in females in one tissue—spleen (Itoh and Arnold 2015). Overall, the report concluded that variability in gene expression was no greater in females than males.

An analogous meta-analysis, using similar methods, was performed focused on neuroscience-relevant measures in male and female rats (Becker et al. 2016). Evaluating over 6000 trait measures, the authors found no overall differences in variability (Fig. 3C). As in prior reports, the analysis identified individual trait categories in which CVs were greater in one sex; and again, this occurred more often in males. Indeed, the authors found that even in the subset of ~300 traits that were measured at known phases of the estrous cycle, segregating data by cycle day did not yield a reduction in variability of female data (Fig. 3D). Taken together, results from meta-analyses of data sets in multiple species and across diverse research domains are characterized by a consistent failure to support the hypothesis that females are intrinsically more variable than males.

Changes in physiology and behavior across the estrous cycle are, however, well-documented in female rodents (Krzych et al. 1978; Barthelemy et al. 2004); indeed, these observations presumably fueled the notion that females must be more variable (and thus more difficult to study). How, then, do these well-established estrous fluctuations fail to result in increased variability in females? A possible answer lies in closer ex-

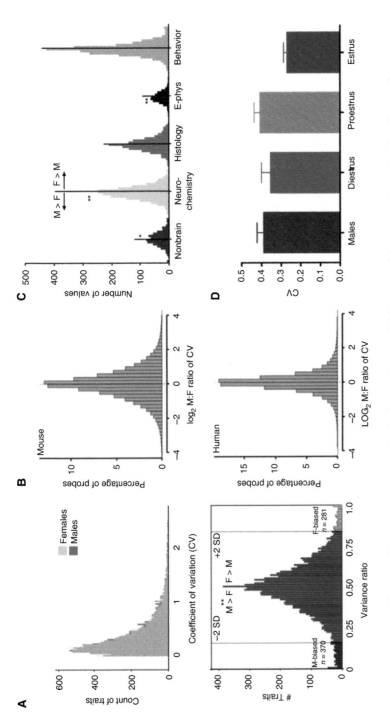

Figure 3. Trait variability is no greater in female than male mice. (*A*) Trait variability comparisons in mice. *Top* panel indicates coefficients of variation (CVs; standard deviation [SD]/mean) for phenotypic traits (*n* = 9932) of male (blue) and female (yellow) wild-type mice as reported in 293 peer-reviewed articles published between 2009 and 2012 (Prendergast et al. 2014). Green shading indicates overlapping areas of yellow and blue histograms. *Bottom* panel depicts CV ratios (CV$_{female}$/[CV$_{female}$ + CV$_{male}$]) for traits in the *top* panel. The CV ratio distribution ranges from 0.0 to 1.0, with CV ratios >0.5 indicating greater trait-specific variance in females, and CV ratios <0.5 indicating greater variance in males. (*B*) CV comparisons in microarray data sets. Histograms depict log$_2$ transformed male-to-female CV ratios across ~2.7 million human and ~2.4 million mouse microarray probes. In most bins, slightly more probes showed CV$_{male}$ > CV$_{female}$ (Itoh and Arnold 2015). (*C*) CV comparisons in laboratory rats. Histogram depicting CV ratios for 6252 traits in laboratory rats, parsed into five broad trait categories. CV ratios were male biased for electrophysiology and neurochemistry measures, and female biased for nonbrain measures. (*D*) Estrous cycle stage did not affect trait variability. Among rats examined in panel C, even when data were aggregated by stage of the estrous cycle, CV$_{female}$ did not differ from CV$_{male}$ (Becker et al. 2016). (Data used under Creative Commons Attribution 4.0 International [CC BY 4.0: https://creativecommons.org/licenses/by/4.0].)

amination of the sources of variance within each sex. Whereas a potent source of female variability may be found in hormonal fluctuations over the ovulatory cycle (Quinlan et al. 2010; Datta et al. 2016), the sources and patterns of male variability have not been systematically evaluated, let alone characterized. Accounting for the overall absence of sex differences in variability is warranted. The magnitude of variance associated with the estrous cycle may be sufficiently small as to have little impact; alternatively, trait

variability of males over several consecutive days may simply be as great as that of females over the estrous cycle. Potential sources of this "hidden" male variance were investigated by performing time-series analyses of locomotor activity and body temperature recorded continuously over the estrous cycle of female mice and in yoked males (Smarr et al. 2017). Overall variability in circadian power was comparable between the sexes for both traits. Remarkably, infradian variability (variance across days) was greater in fe-

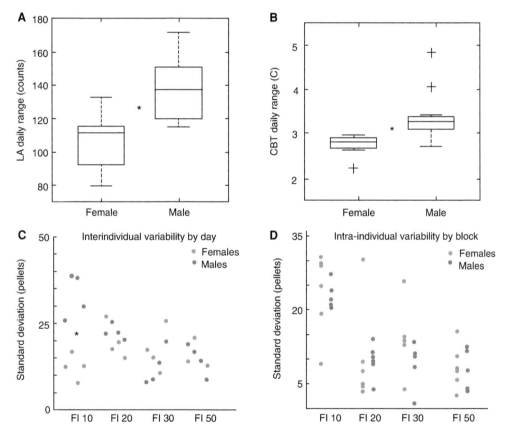

Figure 4. Female mice do not exceed males in variability of activity, body temperature, or food intake. (*A*) Box plots of the daily variability in locomotor activity (LA), and (*B*) core body temperature (CBT) of female and male mice housed in a 12L:12D light:dark cycle indicate that males have a higher intra-animal daily range in LA and CBT. (*C* and *D*) Food intake data within and between male (blue symbols) and female (yellow symbols) mice subjected to schedules of food availability and reward over 4-day blocks (corresponding to the estrous cycle, which was not monitored). Under increasing cost of feeding, females exceeded males in intra-individual variability, (*D*) but not (*C*) interindividual variability. The overall variance did not differ by sex under any of the food reward schedules. Food intake (FI) designations represent the delay in seconds after which responses delivered a 20 mg food pellet. (Data in *A* and *B* are from Smarr et al. 2019; reprinted, with permission, from the authors. Data used under Creative Commons Attribution 4.0 International [CC BY 4.0: https://creativecommons.org/licenses/by/4.0].)

males and ultradian variability (variance within days) was greater in males (Fig. 4A). The results also indicated that, for any given trait, depending on how data are collected (e.g., across multiple days, in a single day, during a fixed time span within a single day), measures will capture variance from different sources. Exclusion of female mice from studies may thus increase variance in investigations in which measures are collected over a span of several hours. Ultimately, this factor may limit generalization of findings from males to females. Smarr and colleagues (2019) also detailed sex differences in variance structure within ingestive behavior. Mice of both sexes were subjected to schedules of food availability and reward over 4-day blocks. Females exceeded males in intra-individual variability, but not interindividual variability (Fig. 4B). Overall, variance did not differ by sex under any of the food reward schedules.

Thus, convergent evidence from multiple species and across diverse traits fails to support the hypothesis that females are more variable than males. Mogil and Chanda (2005) speculated that "male mice [may] feature their own sex-specific variability." Now that the default assumption of greater female variability has been assessed and rejected, a productive new direction may lie in efforts to identify the sources of variance that render males as variable as females, and in many cases, more variable than females.

NEGLECT OF WOMEN IN BIOMEDICAL RESEARCH HAS NEGATIVE HEALTH IMPLICATIONS

Many currently prescribed drugs were approved by the U.S. Food and Drug Administration (FDA) with inadequate enrollment of female animals in preclinical research, and women in clinical trials. For example, the popular sedative-hypnotic drug zolpidem (Ambien), was first approved in 1992 with just 19 women and 49 men (FDA NDA 19908) as the sole pre-approval assessment of the effect of sex on pharmacokinetics (PK), which revealed marked elevation of drug concentrations in women and longer elimination times. Only after decades of post-marketing reports of cognitive deficits in women

were sex-based dose adjustments recommended, at which point women were advised to receive half the zolpidem dose given men. Many other drugs administered in equal doses to women and men require reevaluation for sex-specific dose adjustments, but relevant data are lacking for almost all currently prescribed pharmaceuticals. Zolpidem is but one of many drugs not administered on an mg/kg of body mass basis; women are given the same dose as men, despite lower body weights and sex differences in drug absorption, distribution, bioavailability, metabolism, and excretion. Consequently, women may be overmedicated.

Across all drug categories, ADRs are substantially more common in women than men (de Vries et al. 2019); more female ADR reports were submitted in all regions of the world and all age groups from 12 to 17 years and older (Watson et al. 2019). It is conceivable that some of the increase in ADR reports for women might be caused by the possible overmedication discussed above. Women are also significantly more likely to be hospitalized secondary to an ADR (Tharpe 2011; Nakagawa and Kajiwara 2015; Damien et al. 2016). This disparity is pervasive: 46% of a large sample of drugs manifests significant sex/gender differences in ADRs (Yu et al. 2016). Women over the age of 19 were 43%–69% more likely than men to have an ADR recorded by their general practitioner (Martin et al. 1998). ADRs also peak 20 years earlier among women than men (Martin et al. 1998). Frequently reported ADRs included nausea, headache, drowsiness, depression, excessive weight gain, cognitive deficits, seizures, hallucinations, agitation, and cardiac rhythm anomalies. Women are more likely than men to use two or more medications concurrently (polypharmacy), and more unique medications per year, which may also contribute to increased female ADRs (Manteuffel et al. 2014).

Biological, psychological, and cultural factors contribute to the greater prevalence of ADRs in women, including sex differences in PK and pharmacodynamics (PD), endogenous sex-specific organizational and activational steroid hormone exposure, sex differences in exogenous steroid administration, higher rates of polypharmacy in women, sex differences in the expression of somatoform disorders, and sex differences in reporting rates (Kando et al. 1995).

PK differs in men and women for many drugs (Harris et al. 1995; Meibohm et al. 2002; Schwartz 2003, 2007; Gandhi et al. 2004; Soldin and Mattison 2009; Franconi and Campesi 2014, 2017), which affects drug efficacy and toxicity (Institute of Medicine (US) Committee on Understanding the Biology of Sex and Gender Differences 2001; Amacher 2014). Significant sex differences have been noted in PK in ~28% of bioequivalence studies (Chen 2000). But PK information is included in only a small minority of approved drug labels (Fadiran and Zhang 2015); fewer than 4% of drugs in the *Physicians' Desk Reference* list population PK information in labeling (Duan 2007).

To examine whether sex differences in PK, specifically elevated drug exposure and longer elimination times in women than men, are associated with clinically significant sex differences in ADRs, a literature survey was performed to identify drugs for which sex differences in both PK and ADRs had been examined (Zucker and Prendergast 2020). The analysis identified hundreds of FDA-approved drugs with sex differences in PK, and 86 drugs with statistically significant sex differences in PK, which also reported data on ADRs that were analyzed by sex (Fig. 5). PK differences commonly manifested in measures of the maximum drug concentration in the blood (C_{max}) and area under the curve (AUC), but also included distribution volume and measures of drug elimination rate (e.g., circulating half-life). Many drugs with pronounced sex differences in ADRs were excluded from the analysis because PK data were not available, typically because the drug was approved prior to the year 2000, or because sex-specific analyses were not available in the FDA database. The analysis also indicated that many sex differences in ADRs persisted even after corrections for body weight were considered.

The correspondence of sex differences in PK with sex differences in ADRs was striking. In 88% of instances, a sex difference in PK was linked to a similar sex difference in ADRs; PK-ADR concordance is summarized in Table 1. The sex differ-

ence in ADRs among these drugs is substantially higher than the 46% sex difference in ADRs across all drugs assessed without regard to PK differences (reported by Yu et al. 2016). Thus, including PK data in the consideration of ADRs and stratifying analyses of drugs by the presence and direction of PK differences greatly clarifies patterns of sex differences in ADRs. Nearly all (96%) drugs with higher PK values in women were associated with a higher incidence of ADRs in women than men, whereas only 29% of drugs for which PK values of males exceeded those of females did this sex difference positively predict male-biased ADRs (Fig. 5). Indeed, even in this small fraction of drugs with male-biased PK, ADRs were more prevalent or severe in women. These data show an alarming pattern: elevated drug concentrations and decreased elimination times are far more common in women than men. For drugs with female-biased PK, the clear mapping of PK onto ADRs suggests that drugs that exhibit female-biased PK present a major health risk for women that is not prioritized by the medical profession. An even stronger potential for risk lies in that far larger number of drugs for which no data on PK sex differences exist.

The lack of attention to female subjects during the early stages of drug development may have pervasive, unintended effects that contribute to the disproportionate occurrence of ADRs in women. Drug development pipelines often begin with preclinical modeling, in vitro experiments in human tissues and in vivo experiments in laboratory animal models (usually mice). Inattention to sex and gender in the early stages of drug development can create founder effects that may bias drug efficacy toward one sex. If, during early stages of preclinical development, a drug is optimized and titrated specifically in cells or mice of one sex, then any sex biases inherent in such model systems may be propagated into later stages of drug development. For example, three of the four human cell lines currently being used for SARS-CoV-2 (COVID-19) research are male (Takayama 2020). Disproportionate female-biased PK and ADRs that are likely to emerge may reflect echoes of sex-biased research decisions early in the scientific process.

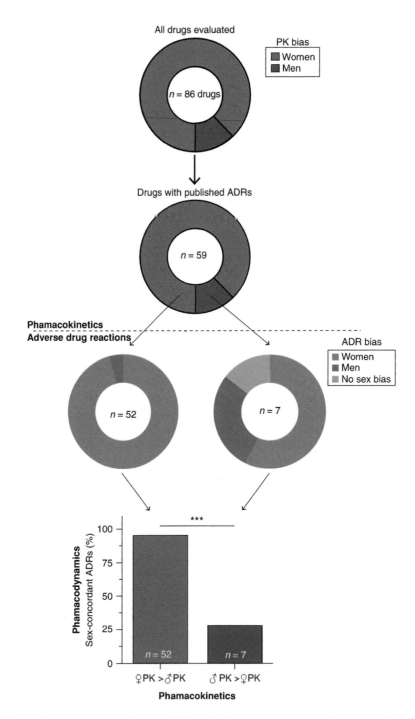

Figure 5. Relations between pharmacokinetics (PK) and adverse drug responses in humans. Sex differences in PK were identified in 86 U.S. Food and Drug Administration (FDA)-approved drugs, 59 of which also yielded data on sex differences in adverse drug reactions (ADRs). PK analyses identified greater drug exposure in women (a female PK bias) for 52 (88%) of these drugs. Among these 52 drugs with female-biased PK, 50 also had female-biased ADRs (96% concordance between PK and ADR biases). In contrast, only two of seven drugs with male-biased PK exhibited male-biased ADRs (29% concordance). (Data from Zucker and Prendergast 2020; reprinted, with permission, from the authors.)

Table 1. Summary of sex-biased pharmacokinetic (PK) measures and concordance with clinical adverse drug reactions (ADRs)

World Health Organization anatomical therapeutic chemical category	Drug[a]	PK measure[b]	Female-biased ADRs	Male-biased ADRs	PK–ADR relation
Alimentary tract, metabolism	Liraglutide	CL_T	Headache, vomiting, nausea	-	Concordant
	Ranitidine	**AUC, T_{max}**		**Duodenal damage**	Concordant
	Rosiglitazone	CL_T	Fracture	-	Concordant
Blood and blood-forming organs	Heparin	CL_T	Blood disorders, lymphatic disorders, bleeding (60 yoa)	-	Concordant
	Aspirin	AUC, CL_T, $t_{1/2}$	Elevated fibrinogen cardiovascular disease in type II diabetics	**Bleeding**	Concordant
	Warfarin	CL_T	Bleeding	-	Concordant
	Clopidogrel	AUC, C_{max}	Fracture, bleeding, gastrointestinal symptoms, inflammatory bowel disease	-	Concordant
Cardiovascular system	Dabigatran	AUC	Bleeding	-	Concordant
	Torasemide	AUC, C_{max}, CL_T, $t_{1/2}$	Hospitalization	-	Concordant
	Pravastatin	AUC, C_{max}	Congenital heart disease	-	Concordant
	Amlodipine	**CL_T**	Edema, flushing, palpitations	-	Discordant
	Digoxin	CL_T	Mortality	-	Concordant
	Verapamil	CL_T	Constipation, edema, fatigue, headache	-	Concordant
	Aliskiren	AUC, C_{max}	Diarrhea	-	Concordant
	Losartan	AUC, C_{max}	-	**Angina, mortality**	Discordant
	Propranolol	AUC, C_{max}, CL_T	Dizziness, muscle pain, headache, dry mouth	-	Concordant
	Dofetilide	SysExp	TdP	-	Concordant
Genito-urinary system, sex hormones	Mirabegron	AUC, C_{max}, SysExp	Treatment discontinuation (CAN), treatment discontinuation (CZE)	**Treatment discontinuation (UK)**	Concordant
	Darifenacin	AUC, C_{max}, CL_T	F > M, ADRs not specified	-	Concordant
	Trospium	AUC, C_{max}	Cognitive impairments	-	Concordant

Continued

Table 1. *Continued*

World Health Organization anatomical therapeutic chemical category	Drug[a]	PK measure[b]	Female-biased ADRs	Male-biased ADRs	PK-ADR relation
Systemic hormonal preparations	Prednisone	AUC, CL_T	Depression, fatigue, hair loss, mood swings, weight gain	-	Concordant
Anti-infectives	Levofloxacin	CL_T, SysExp	Fluoroquinolone toxicity	-	Concordant
	Erythromycin	**CL_T**	TdP	-	Discordant
	Voriconazole	AUC, C_{max}	Cardiac QTc symptoms	-	Concordant
Antineoplastics, immunomodulators	**Cyclosporin**	**CL_T**	-	-	Discordant[c]
	Flurouracil	AUC, CL_T	Stomatitis, leukopenia, alopecia, diarrhea, mucositis	-	Concordant
	Paclitaxel	CL_T	Myocardial infarction, death, lesion, revascularization	-	Concordant
	Capecitabine	AUC	Dose-limiting toxicity	-	Concordant
	Infliximab	CL_T	Allergic reactions	-	Concordant
	Adalimumab	CL_T	Allergic reactions	-	Concordant
Anesthetics, analgesics	Morphine	CL_T	Respiratory depression, hypoxic sensitivity, emesis, nausea	-	Concordant
	Oxycodone	CL_T	Nausea, pruritus, negative affect	-	Concordant
	Buprenorphine	AUC, C_{max}	Sleep disturbance	-	Concordant
	Tramadol	AUC, C_{max}	Treatment discontinuation	-	Concordant
	Zolmitriptan	AUC, C_{max}	Face and neck pressure	-	Concordant
	Ketamine	**CL_T**	-	**Verbal learning, subjective memory**	Concordant
Antiepileptics, anti-Parkinson's	**Carbamazepine**	**CL_T, $t_{1/2}$**	Cognitive impairment, elevated LDL/HDL	-	Discordant
	Gabapentin	C_{max}	Dizziness, somnolescence, nausea	-	Concordant
	Perampanel	AUC, CL_T	Dizziness, headache, treatment, discontinuation	-	Concordant
	Pramipexole	AUC, C_{max}	Nausea, fatigue	-	Concordant

Cite this article as *Cold Spring Harb Perspect Biol* doi: 10.1101/cshperspect.a039156

Psycholeptics	Olanzapine	CL_T	Severe weight gain	-	Concordant
	Clozapine	CL_T	Blood glucose, laxative use, obesity, type II diabetes, neutropenia, leukopenia, weight gain	**Hypertension, increases in BMI, elevated homocysteine, increased basal metabolic rate, increased plasma lipids, QTc symptoms, blood dyscrasias, myocarditis, cardiomyopathy**	Discordant
	Risperidone	CL_T	Hyperprolactinemia, neurological effects, headache hypotension	-	Concordant
	Aripiprazol	AUC, C_{max}	Blood pressure, heart rate, elongated QTc		Concordant
	Diazepam	$t_{1/2}$	Psychomotor impairment	-	Concordant
	Zolpidem	AUC, C_{max}, CL_T	Cognitive impairment, driving impairment	-	Concordant
	Eszopiclone	AUC	Dysgeusia	-	Concordant
	Imipramine	CL_T	Dry mouth, constipation, sweating, tremor, treatment discontinuation	-	Concordant
Psychoanaleptics	Nortriptyline	CL_T	Dry mouth	-	Concordant
	Fluoxetine	CL_T	Hypercortisolemia, elevated albumin, elevated tryptophan, suicidal ideation	-	Concordant
	Citalopram	CL_T	Elevated ADH	-	Concordant
	Sertraline	AUC, CL_T, $t_{1/2}$	Cholesterol, nausea, dizziness, delusions	**Dyspepsia, sexual dysfunction, urinary frequency**	Concordant
	Bupropion	AUC, C_{max}, $t_{1/2}$	EEG abnormalities, seizures	-	Concordant
	Methylphenidate	AUC	Anxiety disorder	-	Discordant

Continued

Table 1. *Continued*

World Health Organization anatomical therapeutic chemical category	Drug[a]	PK measure[b]	Female-biased ADRs	Male-biased ADRs	PK-ADR relation
Antiparasitics	Primaquine	AUC, C_{max}	Nausea	–	Concordant
Respiratory	Terfenadine	C_{max}	TdP	–	Concordant
	Fexofenadine	AUC, C_{max}	Visual attention deficits, reaction time deficits, driving impairments	–	Concordant
Miscellaneous	MDMA	CL_T	Elevated copeptin, hyponatremia	–	Concordant
	Cannabis	CL_T	Increased subjective emotional responses, problematic cannabis use, introvertive anhedonia	–	Concordant

(AUC) Area under the curve, (C_{max}) maximum circulating concentration, (T_{max}) time required to reach C_{max} value, ($t_{1/2}$) elimination half-life, (CL_T) clearance time, exposure, or plasma concentration after a fixed time interval, (SysExp) systemic exposure.
[a]Font style of drug name indicates direction of PK drug exposure bias: (normal font: F > M; **bold font**: M > F).
[b]Font style of PK measure indicates direction of greater drug exposure (normal font: F > M; **bold font**: M > F).
[c]No evidence of sex differences in ADRs.

A number of steps might remediate this sex disparity in health and well-being that stems from female-biased ADRs. The high correlation between elevated PK in women and increased female ADRs suggests that for drugs with higher female PK, the initial dose should be lower for women than men and increased only if the lower dose fails to achieve the desired therapeutic effect.

Establishing sex parity in subject enrollment during the drug approval process should be explicitly identified as a long-term goal of the Department of Health and Human Services. The decades-long pattern of neglect of female animals in preclinical research and underrepresentation of women in clinical trials and research must be corrected; recent NIH oversight and vigilance is an important step in the right direction that needs to be maintained.

CONCLUDING REMARKS

Until relatively recently, female animals and women were woefully underrepresented in biomedical research. Beginning in the 1990s, spurred by the NIH revitalization act and continuing in subsequent decades, a remedial effort has increased inclusion of women and female rodents in clinical and preclinical studies. An increased emphasis on enrolling both sexes emerged in the past decade, but a majority of such studies still fail to consider sex or gender as factors in their analyses; this continued omission represents a missed opportunity and is a serious shortcoming. An increasing number of studies has established that average trait variability is no greater in females than males, thereby removing the long-held, unsubstantiated bias against inclusion of female rodents in research protocols. In many instances, sex inclusion does not require increases in sample size, but when such increases are necessary, they are smaller (e.g., 25% increase) than generally assumed.

The well-established increased susceptibility of women to adverse drug reactions has now been strongly correlated with substantial sex differences in drug PK; women given the same drug dose as men routinely generate higher blood concentrations and longer drug elimina-

tion times and thus may be chronically overmedicated. For drugs with known sex differences in PK, women should be administered lower drug doses.

ACKNOWLEDGMENTS

The authors thank Nicole Woitowich for helpful comments on the manuscript. A.K.B. was supported by NIH Grant R15MH113085 and B.J.P. by NIH Grant AI-67406, the Institute for Mind and Biology and the Social Sciences Division of the University of Chicago.

REFERENCES

Amacher DE. 2014. Female gender as a susceptibility factor for drug-induced liver injury. *Hum Exp Toxicol* **33**: 928–939. doi:10.1177/0960327113512860

Ananthaswamy A, Douglas K. 2018. The origins of the patriarchy. *New Scientist* **238**: 34–35.

Barthelemy M, Gourbal BEF, Gabrion C, Petit G. 2004. Influence of the female sexual cycle on BALB/c mouse calling behaviour during mating. *Naturwissenschaften* **91**: 135–138. doi:10.1007/s00114-004-0501-4

Becker JB, Arnold AP, Berkley KJ, Blaustein JD, Eckel LA, Hampson E, Herman JP, Marts S, Sadee W, Steiner M, et al. 2005. Strategies and methods for research on sex differences in brain and behavior. *Endocrinology* **146**: 1650–1673. doi:10.1210/en.2004-1142

Becker JB, Prendergast BJ, Liang JW. 2016. Female rats are not more variable than male rats: a meta-analysis of neuroscience studies. *Biol Sex Differ* **7**: 34. doi:10.1186/s13293-016-0087-5

Beery AK. 2018. Inclusion of females does not increase variability in rodent research studies. *Curr Opin Behav Sci* **23**: 143–149. doi:10.1016/j.cobeha.2018.06.016

Beery AK, Zucker I. 2011. Sex bias in neuroscience and biomedical research. *Neurosci Biobehav Rev* **35**: 565–572. doi:10.1016/j.neubiorev.2010.07.002

Beltz AM, Beery AK, Becker JB. 2019. Analysis of sex differences in pre-clinical and clinical data sets. *Neuropsychopharmacol* **44**: 2155–2158. doi:10.1038/s41386-019-0524-3

Buch T, Moos K, Ferreira FM, Fröhlich H, Gebhard C, Tresch A. 2019. Benefits of a factorial design focusing on inclusion of female and male animals in one experiment. *J Mol Med* **97**: 871–877. doi:10.1007/s00109-019-01774-0

Chen M. 2000. Pharmacokinetic analysis of bioequivalence trials: implications for sex-related issues in clinical pharmacology and biopharmaceutics. *Clin Pharmacol Ther* **68**: 510–521. doi:10.1067/mcp.2000.111184

Collins LM, Dziak JJ, Kugler KC, Trail JB. 2014. Factorial experiments. *Am J Prev Med* **47**: 498–504. doi:10.1016/j.amepre.2014.06.021

Damien S, Patural H, Trombert-Paviot B, Beyens MN. 2016. Adverse drug reactions in children: 10 years of pharmacovigilance. *Arch Pediatr* **23:** 468–476. doi:10.1016/j.arcped.2016.01.015

Datta S, Samanta D, Sinha P, Chakrabarti N. 2016. Gender features and estrous cycle variations of nocturnal behavior of mice after a single exposure to light at night. *Physiol Behav* **164:** 113–122. doi:10.1016/j.physbeh.2016.05.049

de Vries ST, Denig P, Ekhart C, Burgers JS, Kleefstra N, Mol PGM, van Puijenbroek EP. 2019. Sex differences in adverse drug reactions reported to the National Pharmacovigilance Centre in the Netherlands: an explorative observational study. *Br J Clin Pharmacol* **85:** 1507–1515. doi:10.1111/bcp.13923

Duan JZ. 2007. Applications of population pharmacokinetics in current drug labelling. *J Clin Pharm Ther* **32:** 57–79. doi:10.1111/j.1365-2710.2007.00799.x

Fadiran EO, Zhang L. 2015. Effects of sex differences in the pharmacokinetics of drugs and their impact on the safety of medicines in women. In *Medicines for women* (ed. Harrison-Woolrych M), pp. 41–68. Adis, Cham, Switzerland.

Franconi F, Campesi I. 2014. Pharmacogenomics, pharmacokinetics and pharmacodynamics: interaction with biological differences between men and women. *Br J Pharmacol* **171:** 580–594. doi:10.1111/bph.12362

Franconi F, Campesi I. 2017. Sex impact on biomarkers, pharmacokinetics and pharmacodynamics. *Curr Med Chem* **24:** 2561–2575. doi:10.2174/0929867323666161003124616

Gandhi M, Aweeka F, Greenblatt RM, Blaschke TF. 2004. Sex differences in pharmacokinetics and pharmacodynamics. *Annu Rev Pharmacol Toxicol* **44:** 499–523. doi:10.1146/annurev.pharmtox.44.101802.121453

Harris RZ, Benet LZ, Schwartz JB. 1995. Gender effects in pharmacokinetics and pharmacodynamics. *Drugs* **50:** 222–239. doi:10.2165/00003495-199550020-00003

Heinrich J. 2001. Most drugs withdrawn in recent years had greater health risks for women. General Accounting Office-01-286R, www.gao.gov/products/gao-01-286r

Huang GZ, Woolley CS. 2012. Estradiol acutely suppresses inhibition in the hippocampus through a sex-specific endocannabinoid and mGluR-dependent mechanism. *Neuron* **74:** 801–808. doi:10.1016/j.neuron.2012.03.035

Institute of Medicine (US) Committee on Understanding the Biology of Sex and Gender Differences. 2001. *Exploring the biological contributions to human health. Does sex matter?* National Academies Press, Washington, DC.

Itoh Y, Arnold AP. 2015. Are females more variable than males in gene expression? Meta-analysis of microarray datasets. *Biol Sex Diff* **6:** 18. doi:10.1186/s13293-015-0036-8

Kando JC, Yonkers KA, Cole JO. 1995. Gender as a risk factor for adverse events to medications. *Drugs* **50:** 1–6. doi:10.2165/00003495-199550010-00001

Klein SL, Schiebinger L, Stefanick ML, Cahill L, Danska J, de Vries GJ, Kibbe MR, McCarthy MM, Mogil JS, Woodruff TK, et al. 2015. Opinion: sex inclusion in basic research drives discovery. *Proc Natl Acad Sci* **112:** 5257–5258. doi:10.1073/pnas.1502843112

Krzych U, Strausser HR, Bressler JP, Goldstein AL. 1978. Quantitative differences in immune responses during the various stages of the estrous cycle in female BALB/c mice. *J Immunol* **121:** 1603–1605.

Mamlouk GM, Dorris DM, Barrett LR, Meitzen J. 2020. Sex bias and omission in neuroscience research is influenced by research model and journal, but not reported NIH funding. *Front Neuroendocrinol* **57:** 100835. doi:10.1016/j.yfrne.2020.100835

Manteuffel M, Williams S, Chen W, Verbrugge RR, Pittman DG, Steinkellner A. 2014. Influence of patient sex and gender on medication use, adherence, and prescribing alignment with guidelines. *J Women's Health* **23:** 112–119. doi:10.1089/jwh.2012.3972

Martin RM, Biswas PN, Freemantle SN, Pearce GL, Mann RD. 1998. Age and sex distribution of suspected adverse drug reactions to newly marketed drugs in general practice in England: analysis of 48 cohort studies. *Br J Clin Pharmacol* **46:** 505–511. doi:10.1046/j.1365-2125.1998.00817.x

Mazure CM, Jones DP. 2015. Twenty years and still counting: including women as participants and studying sex and gender in biomedical research. *BMC Women's Health* **15:** 94. doi:10.1186/s12905-015-0251-9

Meibohm B, Beierle I, Derendorf H. 2002. How important are gender differences in pharmacokinetics? *Clin Pharmacokinet* **41:** 329–342. doi:10.2165/00003088-20024 1050-00002

Mogil JS, Chanda ML. 2005. The case for the inclusion of female subjects in basic science studies of pain. *Pain* **117:** 1–5. doi:10.1016/j.pain.2005.06.020

Mogil JS, Wilson SG, Chesler EJ, Rankin AL, Nemmani KVS, Lariviere WR, Groce MK, Wallace MR, Kaplan L, Staud R, et al. 2003. The melanocortin-1 receptor gene mediates female-specific mechanisms of analgesia in mice and humans. *Proc Natl Acad Sci* **100:** 4867–4872. doi:10.1073/pnas.0730053100

Morsink J. 1979. Was Aristotle's biology sexist? *J Hist Biol* **12:** 83–112. doi:10.1007/BF00128136

Nakagawa K, Kajiwara A. 2015. Female sex as a risk factor for adverse drug reactions. *Nihon Rinsho* **73:** 581–585.

Perry MJ, Albee GW. 1998. The deterministic origins of sexism. *Race Gender Class* **5:** 122–135.

Prendergast BJ, Onishi KG, Zucker I. 2014. Female mice liberated for inclusion in neuroscience and biomedical research. *Neurosci Biobehav Rev* **40:** 1–5. doi:10.1016/j.neubiorev.2014.01.001

Quinlan MG, Duncan A, Loiselle C, Graffe N, Brake WG. 2010. Latent inhibition is affected by phase of estrous cycle in female rats. *Brain Cogn* **74:** 244–248. doi:10.1016/j.bandc.2010.08.003

Ramirez FD, Motazedian P, Jung RG, Di Santo P, MacDonald Z, Simard T, Clancy AA, Russo JJ, Welch V, Wells GA. 2017. Sex bias is increasingly prevalent in preclinical cardiovascular research: implications for translational medicine and health equity for women: a systematic assessment of leading cardiovascular journals over a 10-year period. *Circulation* **135:** 625–626. doi:10.1161/CIRCULATIONAHA.116.026668

Rosen R. 1971. Sexism in history or, writing women's history is a tricky business. *J Marriage Family* **33:** 541–544. doi:10.2307/349851

Cite this article as *Cold Spring Harb Perspect Biol* doi: 10.1101/cshperspect.a039156

Schwartz JB. 2003. The influence of sex on pharmacokinetics. *Clin Pharmacokinet* **42:** 107–121. doi:10.2165/00003088-200342020-00001

Schwartz JB. 2007. The current state of knowledge on age, sex, and their interactions on clinical pharmacology. *Clin Pharmacol Therapeutics* **82:** 87–96. doi:10.1038/sj.clpt.6100226

Smarr BL, Grant AD, Zucker I, Prendergast BJ, Kriegsfeld LJ. 2017. Sex differences in variability across timescales in BALB/c mice. *Biol Sex Diff* **8:** 7. doi:10.1186/s13293-016-0125-3

Smarr B, Rowland NE, Zucker I. 2019. Male and female mice show equal variability in food intake across 4-day spans that encompass estrous cycles. *PLoS ONE* **14:** e0218935. doi:10.1371/journal.pone.0218935

Soldin OP, Mattison DR. 2009. Sex differences in pharmacokinetics and pharmacodynamics. *Clin Pharmacokinet* **48:** 143–157. doi:10.2165/00003088-200948030-00001

Sorge RE, Mapplebeck JCS, Rosen S, Beggs S, Taves S, Alexander JK, Martin LJ, Austin J-S, Sotocinal SG, Chen D, et al. 2015. Different immune cells mediate mechanical pain hypersensitivity in male and female mice. *Nat Neurosci* **18:** 1081–1083. doi:10.1038/nn.4053

Takayama K. 2020. In vitro and animal models for SARS-CoV-2 research. *Trends Pharmacol Sci* **41:** 513–517. doi:10.1016/j.tips.2020.05.005

Tharpe N. 2011. Adverse drug reactions in women's health care. *J Midwifery Womens Health* **56:** 205–213. doi:10.1111/j.1542-2011.2010.00050.x

Wald C, Wu C. 2010. Biomedical research. Of mice and women: the bias in animal models. *Science* **327:** 1571–1572. doi:10.1126/science.327.5973.1571

Watson S, Caster O, Rochon PA, den Ruijter H. 2019. Reported adverse drug reactions in women and men: aggregated evidence from globally collected individual case reports during half a century. *EClinicalMedicine* **17:** 100188. doi:10.1016/j.eclinm.2019.10.001

Woitowich NC, Woodruff TK. 2019. Implementation of the NIH sex-inclusion policy: attitudes and opinions of study section members. *J Womens Health* **28:** 9–16. doi:10.1089/jwh.2018.7396

Woitowich NC, Beery A, Woodruff T. 2020. A 10-year follow-up study of sex inclusion in the biological sciences. *eLife* **9:** e56344. doi:10.7554/eLife.56344

Yu Y, Chen J, Li D, Wang L, Wang W, Liu H. 2016. Systematic analysis of adverse event reports for sex differences in adverse drug events. *Sci Rep* **6:** 24955. doi:10.1038/srep24955

Zajitschek SR, Zajitschek F, Bonduriansky R, Brooks RC, Cornwell W, Falster DS, Lagisz M, Mason J, Senior AM, Noble DW, et al. 2020. Sexual dimorphism in trait variability and its eco-evolutionary and statistical implications. *eLife* **9:** e63170. doi:10.7554/eLife.63170

Zucker I, Prendergast BJ. 2020. Sex differences in pharmacokinetics predict adverse drug reactions in women. *Biol Sex Diff* **11:** 1–14. doi:10.1186/s13293-020-00308-5

Quantifying Sex Differences in Behavior in the Era of "Big" Data

Brian C. Trainor[1] and Annegret L. Falkner[2]

[1]Department of Psychology, University of California, Davis, California 95616, USA

[2]Princeton Neuroscience Institute, Princeton, New Jersey 08540, USA

Correspondence: bctrainor@ucdavis.edu

Sex differences are commonly observed in behaviors that are closely linked to adaptive function, but sex differences can also be observed in behavioral "building blocks" such as locomotor activity and reward processing. Modern neuroscientific inquiry, in pursuit of generalizable principles of functioning across sexes, has often ignored these more subtle sex differences in behavioral building blocks that may result from differences in these behavioral building blocks. A frequent assumption is that there is a default (often male) way to perform a behavior. This approach misses fundamental drivers of individual variability within and between sexes. Incomplete behavioral descriptions of both sexes can lead to an overreliance on reduced "single-variable" readouts of complex behaviors, the design of which may be based on male-biased samples. Here, we advocate that the incorporation of new machine-learning tools for collecting and analyzing multimodal "big behavior" data allows for a more holistic and richer approach to the quantification of behavior in both sexes. These new tools make behavioral description more robust and replicable across laboratories and species, and may open up new lines of neuroscientific inquiry by facilitating the discovery of novel behavioral states. Having more accurate measures of behavioral diversity in males and females could serve as a hypothesis generator for where and when we should look in the brain for meaningful neural differences.

SEX DIFFERENCES IN BEHAVIOR MAY BE SIMULTANEOUSLY OVERLOOKED AND OVEREMPHASIZED

Early in the study of animal behavior, dramatic sex differences observed in courtship and mate selection behaviors launched the idea that behaviors themselves were under evolutionary control, and enforced long-held assumptions that there might be "male" behaviors and "female" behaviors (Darwin 1882). We now know that sex differences in investment in reproductive and mating strategies can lead to clear and quantifiable sex differences in behavior that directly affect the fitness of an individual. Historically, the study of sex differences focused on the most extreme differences in behavior including reproductive behaviors (Ball et al. 2014) and aggression (Lischinsky and Lin 2020). However, the absence of an obvious link to parental or reproductive strategy does not mean that there are not important sex differences in behavior, let alone in the underlying neural dynamics or circuit architecture that generate the behavior (De Vries and Boyle 1998). Although it may be obvious to look for sex differences in social behaviors that are directly related to mating and aggression, there is growing evidence that subtle sex differences may have been systematically overlooked.

In contrast, although some sex differences in behavior may have been ignored through a hyperfocus on behaviors with direct links to reproduction, described sex differences in behaviors that "do" have direct links to reproductive strategy may, in some cases, be overblown. For example, in many species males but not females use vocalizations to attract mates or compete with other individuals. There has been a recent appreciation that in some species females also use vocalizations in competitive interactions (Kelly 1993; Emerson and Boyd 1999). Thus, descriptions of some behaviors as sexually dimorphic may be reinforced by existing biases about "male" and "female" behaviors, and deserve further scrutiny.

Implementing more complete descriptions of behavior for both sexes is an important remedy for overcoming these biases. We frequently lack complete behavioral descriptions for both sexes as, for many years, females were not included in behavioral analyses due either to neglect or misguided beliefs. Females were often excluded from studies because of the concern that hormone fluctuations across the reproductive cycle would obscure the neuroscientific objectives. This rationale has been discredited (Prendergast et al. 2014; Becker et al. 2016) and the consideration of "sex as a biological variable" (SABV) by the NIH has been a critical and necessary corrective to a decades-long imbalance in neuroscience (but see Mamlouk et al. 2020). When this more balanced approach is applied to more foundational behavioral "building blocks" such as reward learning or defensive behavior, subtle sex differences in these behaviors are sometimes found hiding in plain sight.

One example of a "missed" behavior is the recent finding that female rats may use a different behavioral strategy than male rats during the expression of learned fear. Historically, learned fear has been quantified by observing canonical freezing behavior. In many fear conditioning studies females show reduced freezing relative to their male counterparts (Fig. 1A; Maren et al. 1994; Pryce et al. 1999; Clark et al. 2019). However, through careful analysis of video recordings, researchers identified a new behavior —"darting"—that was far more prevalent in females than in males (Fig. 1B; Gruene et al. 2015).

Why does this matter? If a field, as a result of several decades of exclusive research on the behavior of males, has been focused on freezing as the de facto defensive reaction, then the responses of the females, who freeze less often, appear aberrant. The fact that females use more diverse strategies to defend against threat should prompt a rethink about what it means when an animal freezes (or does not freeze) and how this behavioral diversity should be taken into account when assessing how animals learn. The identification of darting behavior was in some ways accidental. It was the product of taking a wider view of which behaviors individuals of both sexes performed during a routine experimental paradigm: fear conditioning. The accidental nature of this discovery raises the question of whether a more systematic approach can be used to evaluate whether there are important sex differences in other behaviors. As we enter a new era in the ability to collect and analyze behavior data with more sophistication and quantitative rigor, it may be time to ask how we can best harness new technology to discover subtle, but potentially important, patterns in behavior and reduce our reliance on single-variable readouts for internal states.

Several recent reviews have called for a return to a more ethologically motivated study of behavior in neuroscience (Anderson and Perona 2014; Krakauer et al. 2017; Datta et al. 2019). Implicit in these arguments is that there is new knowledge to be gained by moving the lens away from the brain and instead becoming more focused on behavior, and that insights about behavior can be used to direct neuroscientific inquiry. However, many of the examples used in these calls to action do not explicitly consider behavior in males and females. To take a more holistic look at behavior, individual differences

   Cite this article as *Cold Spring Harb Perspect Biol* doi: 10.1101/cshperspect.a039164

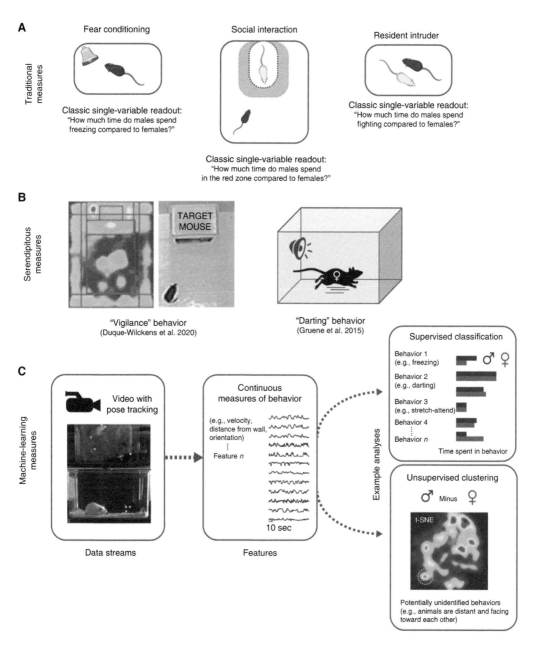

Figure 1. New tools to facilitate the discovery of sex differences in behavior. (*A*) Common behavioral paradigms used in neuroscience commonly rely on single-variable descriptions of complex behaviors for both sexes. Examples include fear conditioning (*left*), social avoidance (*center*), and the "resident intruder" paradigm for social behavior (*right*). (*B*) Recent examples of serendipitously discovered behaviors in these paradigms suggest that single variables are insufficient to describe sex differences in behavior. Examples include increased social vigilance in females (*left*), and increased "darting" behaviors in females during fear conditioning (*right*). (*C*) Adopting machine-learning strategies for pose tracking and video analysis will facilitate behavioral discovery by allowing the robust quantification of multiple behaviors, and may enable the discovery of new behaviors. One example is using video acquisition and pose tracking (in this case, DeepLabCut), to reveal behavioral "features" from supervised analyses (*right top*), but also applying unsupervised analyses (*right bottom*) to provide a more nuanced view of sex differences in complex behavior. Difference heat map may reveal behavior clusters that are enriched in one sex over the other.

across both sexes must be measured, such that appropriate behavioral readouts that fully represent the behavior of both sexes can be designed.

Single-variable measurements for complex internal states abound in the scientific literature. Initial developments in the computerization of behavioral analysis automated a narrow set of measurements allowing high-throughput behavioral analyses. However, simply measuring time in the center of an open field, or time spent in an interaction zone during a social interaction test, can overlook other important behavioral patterns (von Ziegler et al. 2021). Developments in machine vision and machine learning provide an opportunity to apply neuroethological approaches without losing the throughput of automated behavioral scoring. Often behavioral analysis is used to infer an underlying state such as fear or motivation. Using more complete descriptions of behavior in both sexes under ethologically relevant conditions is likely to provide richer insights into these states. The reasons for this are simple: Animals of different sizes, physiology, or life experience might apply different strategies for deciding which behaviors to use and when. Using the same logic for why we might use age-matched cohorts to study a particular behavior (because age-matched cohorts should be of comparable size and experience), we can extend this same logic to looking at sex differences in a broad range of behaviors. The use of machine learning confers additional advantages, since this approach may enable the detection of behaviors or behavioral sequences that would be difficult or impossible for humans to detect. Historically, behaviors have been classified by experts with deep knowledge of the species being studied. However, even an expert could be biased by the way the human sensory system perceives motor patterns. The development of novel algorithms that detect or classify behavior with minimal or no human intervention could provide more unbiased descriptions of behavior and potentially even identify previously undescribed behaviors missed by experts.

Here, we will survey a brief history of general approaches to quantifying sex differences in behavior, and consider new directions for the field as it incorporates new computational tools to simplify increasingly high-dimensional and multimodal data. In addition, we will propose a model of how researchers might use this information to guide neural circuit interrogation of how these behavioral differences are enforced.

QUANTIFYING SEX DIFFERENCES WITHOUT BIAS

Then and Now: A Brief History of Quantifying Behavioral Difference

Studying sex differences in behavior has deep roots in neuroethology, because some of the first well-described behaviors, including egg-protecting maneuvers famously described by Karl Lorenz, are sexually dimorphic. These discoveries were based on human observations performed either directly or through recordings. Several methodologies were designed to account for the limits of human abilities (Martin and Bateson 1994). To obtain more detailed observations, an investigator could use "ad libitum" sampling and note down whatever is visible or relevant during an observation period. This approach provides flexibility to record unusual behaviors, but is prone to overlooking subtle responses, especially in a group setting (Bernstein 1991).

An alternate approach is scan sampling, in which individuals are observed at brief, regular intervals and the observer uses a defined list of behaviors to record. Scan sampling allows for the monitoring of multiple individuals but at limited depth. Furthermore, the brief scans will usually miss infrequent behaviors. Focal sampling, in which the investigator focuses observations on a single individual, is more conducible to detecting rare behaviors. Similar to scan sampling, the behaviors to be scored are decided in advance, and in laboratory settings the focal animal is usually visible for the entire observation period. If combined with video recordings, it is possible for an investigator to intensively observe behavior from multiple individuals interacting at the same time. However, this approach is laborious, as the investigator must score the same recording multiple times, each time focusing on a different individual. Although it is usually possible for multiple observers to be trained to score behav-

   Cite this article as *Cold Spring Harb Perspect Biol* doi: 10.1101/cshperspect.a039164

ior with high reliability, there is a limit to what can be realistically expected from human observers. In addition, some interactions between individuals may occur synchronously and may not be readily identified by observers focused on individual actions rather than those of a group or pair. Thus, even heroic efforts on the part experimenters using existing methods are likely to miss important behavioral sex differences that could potentially provide significant insights into observed neural differences. The adoption of semi-automated and automated methods for behavioral capture, including behavioral "pose" tracking, could add a new level of rigor to behavioral analyses that can identify new behavior patterns or overlooked sex differences.

Semi-Automated and Automated Methods

One of the first forms of automated behavioral tracking was computerized tracking of beam breaking as a measure of general locomotor "activity." This allows monitoring of more individuals over a longer period of time. One example of a sex difference in locomotor behavior is seen in response to pharmacological compounds. Female rodents show an exaggerated locomotor response to cocaine compared to males (Roth and Carroll 2004; Van Swearingen et al. 2013), a difference that appears to be established before birth (Forgie and Stewart 1993). The development of commercial video tracking systems led to more widespread use of computerized analysis of behavior as it allows more flexibility in the types of apparatuses that can be used and for more parameters to be quantified. One useful variable is path length, which is laborious to calculate by eye and provides a more accurate measure of search strategy in learning and memory tests such as the water maze. For example, although male rats took shorter paths to reach the hidden platform than females, this difference was eliminated if rats were given an initial training session to familiarize them with the requirements of the water maze tests (Perrot-Sinal et al. 1996). One drawback to video tracking methods is that subtle or complex behaviors can be difficult to detect when data collection is limited to x, y coordinates.

Although human-driven behavioral annotation of more subtle or complex behaviors remains a staple of behavioral analysis, the ability to speed up this process has been assisted by several pieces of software that facilitate video "tagging," including open source options like the Behavioral Observation Research Interactive software (Friard and Gamba 2016). These software suites enable users to examine video frame-by-frame and "tag" relevant behavioral moments (e.g., the onsets and offsets of user-defined behaviors). However, for aggressive behaviors, there is significant variability in how individuals score behaviors (Segalin et al. 2020). Furthermore, this approach is time consuming and relies on behavior patterns defined by the experimenter, which themselves may be biased. Thus, these methods may be less conducive for discovering novel behaviors or sex differences.

A significant advance in our ability to quantify complex behavior arrived in the form of pose tracking, in which specific body parts are tracked to define behavioral patterns. Although tracking the pose of individuals is not a new idea (Marr and Nishihara 1978), the ability to do it reliably, inexpensively, and with limited computational knowledge, is novel. Motivated by significant recent advances in automated video analysis, including the use of trained "deep" convolutional neural networks to reliably identify the content of an image, these methods have revolutionized the ability of researchers to study animal behavior in the laboratory and beyond (Graving et al. 2019; Li et al. 2020). In particular, machine-learning-based methods, including user-friendly software such as DeepPoseKit, LEAP, DeepLabCut, or MARS, allow users to perform automated detection of user-defined points (e.g., the forelimb or nose of a mouse) across frames of a given video data set (Mathis et al. 2018; Graving et al. 2019; Pereira et al. 2019; Segalin et al. 2020). These algorithms rely on a user-generated training data set, in which the user labels relevant body parts (e.g., an animal's forelimb or nose) in a subset of videos to train the algorithm. After training, the algorithm can then detect the presence or absence of those selected features in the remaining data set. Depending on the video quality and the discriminability of the features being tracked, pose tracking can be implemented with little or no algorithmic "proofreading," the process of iteratively refining the model. Importantly, the use of these methods is scalable. With previous methods, manually tracking multiple features or even multiple animals significantly increases the time cost to the experimenter. However, the use of automated pose-tracking approaches allows the application of large numbers of labels to an ever-expanding video data set. An important point to note is that pose tracking, which ultimately results in a time series of spatial locations of desired body parts, may not directly yield usable behavioral data. Instead, these spatial locations are often first converted into relevant features that represent specific relationships between individual points (e.g., the nose-to-nose distance between two interacting rodents). To extract additional meaningful information from pose-tracking data, either supervised or unsupervised classification strategies based on the fields of statistics and machine learning can be applied. Several excellent recent reviews thoroughly outline how these methods may be used to quantify behavioral dynamics (Mathis et al. 2020; Pereira et al. 2020), but we will summarize their use briefly here.

Supervised strategies, including behavioral classification, are frequently used to identify instances of a behavior in a larger data set. These strategies rely on a user-generated ground truth, in which an expert labels instances of a user-specified behavior (e.g., are mice fighting in this frame?) to train the algorithm to link specific poses with these behaviors on a frame-by-frame basis. Recent examples of this method used support vector machines and random forest classification to assist in identifying instances of user-defined behaviors from pose-tracking data (Nilsson et al. 2020). A strength of this method is that, because the user defines the behaviors to be tracked, the end data are straightforward to interpret. For example, the frequency, bout length, or duration of a particular behavior can be compared in two groups. However, this strength is also a constraint: detected behaviors are limited to what the user has already specified as being important, which precludes the ability to identify novel behavioral patterns. For exam-

ple, some sex differences in behavior may lie outside our expectations (e.g., darting behavior). A second drawback to these strategies is that they perform poorly in situations in which data are limited or behaviors of interest are extremely rare or ambiguously defined by the user. Therefore, additional methods may need to be used that allow for the identification of "new" behaviors.

Unsupervised methods, which infer the structure of the data without ground truth labeled outputs will likely yield additional insights. Although supervised methods excel at extracting patterns to assist in identifying the "known unknowns," unsupervised methods can be used in parallel to identify behaviors that might have been missed by experimenters (the "unknown unknowns"), or to identify latent structure in the data. Successful examples of this approach are beginning to emerge using pose-tracking data as the input. For example, using a nonlinear embedding approach on pose-tracking data from freely moving individuals revealed interpretable behavioral "clusters" (Berman et al. 2014; Klibaite et al. 2017), in which each cluster represented an identified behavior (e.g., "turning left" or "grooming"). Other clusters may represent behavioral states (e.g., two individuals interacting socially are at a specific distance and orientation relative to each other) that may not, to the human observer, initially appear to be distinct "behaviors." However, further analysis could prove that motor patterns identified by unsupervised analyses may, in fact, be informative about the internal state of the animals or how they pattern their next choice of behavior. Open source tool kits, including B-SOiD, have made unsupervised data exploration more tractable for the end user (Hsu and Yttri 2021). In head-fixed preparations for imaging and electrophysiology, high-speed video of the face and body is often collected alongside neural data. Significant progress has been made using supervised and unsupervised approaches to detect signatures of facial expression or emotional state (Khan et al. 2020; Dolensek et al. 2020), which previously could only be identified with painstaking frame-by-frame analysis by human observers (Grill and Norgren 1978). However, even with these automated methods, sex differences are rarely reported. Given the strong evidence for sex differences in physiological and behavioral responses to stress (Laman-Maharg and Trainor 2017), it seems likely that males and females could differ in the use of facial expressions.

Although these tools provide new ways to discover new behaviors, scientists with strong training in behavioral analyses are still needed to assess the importance of motor patterns identified by unsupervised methods and design experiments to determine their function.

INCORPORATING MULTIMODAL DATA STREAMS

Of course, not all behavior is detectable bodily movement. Other multimodal data streams can be integrated with pose tracking to create a more holistic picture of behavior. Some forms of communication, for example, are nearly impossible to detect through limb tracking (although they might be related). For many rodent researchers, recording ultrasonic vocalizations (USVs) has revealed critical insights about courtship, parental, and prosocial behaviors (Holy and Guo 2005). In addition to auditory streams of data, other important behavioral variables, including changes in body temperature (using temperature-sensitive cameras) or odor profile (using a photoionization detector) can be integrated with ongoing movement detection to provide additional behavioral context.

In addition to their use in video data, supervised and unsupervised strategies have also been successfully applied to the detection and segmentation of "big" audio data, in particular to the identification of rodent calls (Van Segbroeck et al. 2017; Coffey et al. 2019; Vogel et al. 2019). For communication signals emitted in groups of animals, experimenters have to solve two specific problems: (1) What is the call being emitted (call classification), and (2) Who is making this call (call identification)? These problems pose unique challenges because audio data is frequently collected from multiple spatially separated microphones, and because calls are made during social interactions, communication signals are often overlapping from multiple individuals, making it difficult to determine which individual is vocalizing.

For call classification, using a combination of manual call identification, supervised classified calls, and unsupervised classified calls, the full repertoire of rodent communication signals is starting to emerge. Although some algorithms for USV segmentation (e.g., DeepSqueak and MUPET) use the acoustic features of the audio signal to classify or segment the sound, use of variational autoencoders to define a latent space (a lower dimensional representation of the highly complex acoustic spectrogram) can be used to directly compare these models (Goffinet et al. 2021). These models have converged on a common set of vocal syllables that make up the full mouse vocal repertoire, which allow the content of calls to be compared between males and females. The ability to classify the content of calls can now be combined with positional information or pose-tracking data captured by video to allow recorded calls to be "assigned" to an individual. These data, combined with behavioral analysis described above, further allows additional insights about which behaviors are correlated with specific vocalizations.

Vocal communication in rodents and birds has long been believed to be strongly sexually dimorphic, with females vocalizing rarely, if ever. However, use of more unbiased methods for audio analysis combined with positional information has revealed that female mice, rather than being silent recipients of male communicative signals, are active participants (Neunuebel et al. 2015), in some cases contributing nearly 20% of the total vocalizations. A detailed analysis of the structure of female calls shows that they are largely similar to the calls of the male (Hammerschmidt et al. 2012), but are more likely to call to other females than to males. This second look at the prevalence of female song in rodents is echoed in recent work in birds, which suggests that birdsong in many species is not only prevalent in females, it is conserved across species (Odom et al. 2014). This finding represents a good example of how use of an unbiased behavioral lens to look at sex differences decreases the size of an assumed sex difference.

AUTOMATED BEHAVIORAL PHENOTYPING TO ASSESS INTERNAL AND AFFECTIVE STATES

The use of machine-learning approaches to behavioral analysis can be used not only to assess differences in moment-to-moment behaviors, but also to make predictions about an individual's internal or affective state. Sex differences in these metrics of well-being have already been observed using automated and semiautomated methods. For example, acute behavioral phenotyping in the home cage shows sex differences in the stress responses of mice, in which females show more depression-like phenotypes than males such as reduced grooming and general activity (Goodwill et al. 2019). These effects were reversed by antidepressant treatment and echoed more traditional tests of rodent depression-related behavior such as sucrose anhedonia. In a more long-term experiment, full-time "24/7" monitoring of mice of both sexes in their home cages across long timescales reveals sex differences in how individuals respond to a cage change stress, with males showing a more variable stress response to cage changing (Pernold et al. 2019). These methods can also be used to further quantify individual variability both between and within sexes. Such "behavioral phenotyping" has been recently used in fish and mice to quantify individual differences in behavior in response to various pharmacological perturbations (Hoffman et al. 2016; Wiltschko et al. 2020) and can be further extended to look at sex differences.

Understanding sex differences in the way individuals respond to stress is critical for developing behavioral methods to evaluate affective state. In several species of rodents, both males and females show strong approach responses to an unfamiliar social context (Duque-Wilckens et al. 2016; Yohn et al. 2019). The most widely used method to measure this response is a social interaction test (Fig. 1A) in which an unfamiliar target mouse is placed into a small wire cage and the time the focal mouse spends near this cage is quantified (Golden et al. 2011). This single approach to capturing the stress response is likely insufficient to capture behavioral variability across sexes. Exposure to social stressors (in the

form of aggressive interactions) can induce different behavioral phenotypes in males and females. In California mice (*Peromyscus californicus*), three brief episodes of social stress exposure reduces social approach in females but not males (Trainor et al. 2011, 2013). If a large arena is used for testing, additional behaviors can be observed. When there is more space, stressed females orient toward the target mouse while simultaneously avoiding it, a response referred to as social vigilance (Duque-Wilckens et al. 2018). Social vigilance appears to function as a risk-assessment behavior in adverse or changing social environments (Wright et al. 2020).

Like darting behavior, social vigilance was largely discovered by accident. Heat maps produced from social interaction tests indicated that many stressed females would spend large amounts of time in the center of the arena (Fig. 1B; Greenberg et al. 2014). Initially this finding was baffling, because anxiety-like states are associated with decreased time spent in the center of an open field. Only after measuring several different variables produced by computer tracking software was it discovered that stressed females were orienting to a stimulus mouse while simultaneously avoiding it (Fig. 1B). In California mice, social stress exposure induces an enduring increase in social vigilance in females but not males 2 weeks later. The neural circuitry for social vigilance is intact in male California mice, as it is observed immediately after an episode of defeat. However, this effect is transitory (Duque-Wilckens et al. 2020). Social vigilance has also been observed in stress models using C57Bl/6J mice. In one approach, sexually experienced CFW females reliably showed aggression toward female C57Bl/6J in a resident-intruder test (Newman et al. 2019). C57/Bl6J females exposed to 10 such episodes of defeat showed reduced social contact in the home cage and increased social vigilance in an unfamiliar arena. In the unfamiliar arena, inclusion of social vigilance was important because standard measures of time spent in an interaction zone did not differ between control and stressed females. In an alternative approach, male or female C57Bl/6J observes another male experience social defeat (Warren et al. 2020). This vicarious exposure to defeat stress induces social

Cite this article as *Cold Spring Harb Perspect Biol* doi: 10.1101/cshperspect.a039164

vigilance in both male and female C57Bl/6J, but the effect is significantly stronger in females (Duque-Wilckens et al. 2020).

This serendipitous discovery of social vigilance suggests that more holistic, quantitative descriptions of behavior could generate significantly more insight than the standard single-variable measure of affective state or stress response (e.g., measures of time spent in an interaction zone). Machine-learning approaches could open new doors for identifying important sex differences in behavioral responses to stress. Because paradigms including chronic social defeat and social interaction tests have been widely applied, standardization across laboratories in how behaviors are quantified and scored could generate new opportunities for generating large data sets that could be directly compared. For example, if video recordings were standardized, pose tracking could be used in a consistent way across laboratories and fed into either supervised or unsupervised algorithms to identify specific behavioral patterns that differ between sexes (Fig. 1C). Although social approach and vigilance can be quantified using comparatively simple video tracking software that tracks both the head and the base of the tail, other behaviors may be subtle. For example, rodents respond to both predators (Hubbard et al. 2004) and social threats (McCann and Huhman 2012) with a "stretch-attend" posture in which the individual crouches to reduce visibility and slowly approaches the threat. Stretch-attend is traditionally scored by trained observers and is more easily quantified from a side view versus the overhead view that is more typically used in social interaction tests. Deep-learning approaches that track multiple body points (Mathis et al. 2018; Pereira et al. 2019), combined with machine-learning tools (Sturman et al. 2020), are better suited to detecting these posture-specific behavior patterns than standardized video tracking systems.

MAPPING SEX DIFFERENCES IN BEHAVIOR TO SEX DIFFERENCES IN THE NERVOUS SYSTEM

Comprehensive, quantitative descriptions of behavior can generate hypothesis for the neural mechanisms that underlie putative differences. For example, sex differences in behavior can occur on a spectrum from completely binary in a single behavior, to extremely subtle and across a diverse set of behaviors. Knowledge about effect sizes across a range of behaviors may give clues to the neural mechanism underlying the difference.

When behaviors appear to have a bimodal distribution, that may indicate that one sex completely lacks the ability to perform this behavior, and clear sexual dimorphism might be observed in motor processing circuits. This is the case in some species of frogs in which males but not females use advertisement vocalizations to attract females and compete with other males. In African clawed frogs, the motor neurons and muscle fibers that control vocalizations are androgen-sensitive, and increased androgen levels during puberty are essential for normal development of the laryngeal muscles required for vocalizations (Tobias et al. 1993; Emerson and Boyd 1999). However, careful analysis revealed that in some species of frogs, females also produce advertisement calls (Kelly 1993; Emerson and Boyd 1999). The mechanisms for advertisement calls in females have been understudied (Wilczynski et al. 2017). A potentially important neuropeptide is arginine vasotocin, which can induce advertisement calls in female tree frogs even in the absence of androgen treatment (Penna et al. 1992). Although these calls have different frequencies than those of males, this example shows the importance of considering both sexes, even in cases in which behavior at first glance appear to be strongly sexually dimorphic.

However, for most sex differences in behavior there is continuous variability in frequency, intensity, or duration. One example is aggression. Despite the fact that in most species, males are more aggressive than females, this appears to be a matter of degree and not kind. Although males are in fact more aggressive in most species, female aggression is often displayed in different contexts. Instead, female aggression is most frequently displayed during lactation. Key brain areas for patterning aggressive motivation and action, including the ventromedial hypothalamus, ventrolateral area (VMHvl), and the periaqueductal gray (PAG), appear to be largely similar in males and females (Lin et al. 2011; Hashikawa et al. 2017; Falkner et al. 2020). Rather than showing large structural differences, aggression circuits instead appear to differ in the number of neurons that are coactivated by same- and opposite-sex social partners. Although males have a "shared" population of VMHvl neurons that responds to both male and female conspecifics, VMHvl neurons in females that respond to males and females are functionally and anatomically separated (Falkner et al. 2014; Hashikawa et al. 2017). These subtle, yet critical differences may underlie sex differences in when and how aggression is deployed by changing how sensory information is integrated within these circuits.

There can also be sex differences in how neural circuits respond to the environment, rather than in the neural circuits that directly produce behavior. For example, analyses of neural circuits modulating social approach in California mice reveal few sex differences. In unstressed males and females, oxytocin receptors within the nucleus accumbens (Williams et al. 2020) and vasopressin V1a receptors in the bed nucleus of the stria terminalis (BNST) promote social approach. Indeed, across numerous analyses of gene expression or neuropeptides in the nucleus accumbens (Campi et al. 2014) and BNST (Greenberg et al. 2014; Duque-Wilckens et al. 2016, 2018; Steinman et al. 2016), few sex differences are observed. However social defeat induces sex-specific effects in these circuits (Steinman et al. 2019). For example, social defeat increases oxytocin synthesis as well as the reactivity of oxytocin neurons within the BNST (Steinman et al. 2016). Oxytocin synthesis in the BNST is necessary for stress-induced vigilance in female California mice, whereas oxytocin infusions into the BNST are sufficient to increase vigilance and reduce social approach in both males and females (Duque-Wilckens et al. 2020). Thus, it appears that a sex difference in the effects of stress on oxytocin neurons within the BNST drives sex differences in social approach and vigilance. Sex differences in transcriptional responses to social defeat have also been observed in the nucleus accumbens (Hodes et al. 2015; AV Williams et al., in prep.).

These sex-specific effects of stress on transcription within the brain contribute to important sex differences in behavioral responses following stress.

Beyond looking for "which" brain regions might have a role in mediating sex differences, comprehensive behavioral phenotyping will also allow us to assess "when" sex differences in behavior occurs. Sex differences in behavior may be seasonal (Henningsen et al. 2016) or may have different trajectories of expression across the life span (Schulz and Sisk 2016; Choleris et al. 2018). Collecting video data at different time points is now relatively low cost, and may reveal new insights about development.

CONCLUSION: SEX DIFFERENCES ARE "INDIVIDUAL" DIFFERENCES

The arc of neuroscience research is starting to bend back toward behavior. The development and distribution of new tools for animal pose tracking, in conjunction with the ability to record and integrate other multimodal streams of information, now allow us to capture a more holistic picture of animal behavior in the laboratory than has ever been possible. Broad access to these tools and the adoption of evolving strategies for analyzing these data may propel the next generation of neuroscientific researchers to fully characterize behavior, including sex differences in behavior, before launching a study. Not only will this approach yield novel behavioral insights, enabling the discovery of new behaviors and internal states, it will prevent potential experimental pitfalls, including oversimplistic single-variable readouts for complex states. Last, understanding the richness of behavioral variability (individual differences) is a long sought-after experimental goal. The quest to understand sex differences in behavior should be framed in this light, rather than in the rather narrow and often inaccurate confines of "maleness" and "femaleness." Striving to constantly and thoroughly evaluate population-wide individual variability in behavior will pay dividends in the ability to interpret neuroscientific data, and to design insightful behavioral readouts of complex internal and affective states.

REFERENCES

Anderson DJ, Perona P. 2014. Toward a science of computational ethology. *Neuron* **84:** 18–31. doi:10.1016/j.neuron.2014.09.005

Ball GF, Balthazart J, McCarthy MM. 2014. Is it useful to view the brain as a secondary sexual characteristic? *Neurosci Biobehav Rev* **46:** 628–638. doi:10.1016/j.neubiorev.2014.08.009

Becker JB, Prendergast BJ, Liang JW. 2016. Female rats are not more variable than male rats: a meta-analysis of neuroscience studies. *Biol Sex Differ* **7:** 34. doi:10.1186/s13293-016-0087-5

Berman GJ, Choi DM, Bialek W, Shaevitz JW. 2014. Mapping the stereotyped behaviour of freely moving fruit flies. *J R Soc Interface* **11:** 20140672. doi:10.1098/rsif.2014.0672

Bernstein IS. 1991. An empirical comparison of focal and ad libitum scoring with commentary on instantaneous scans, all occurrence and one-zero techniques. *Anim Behav* **42:** 721–728. doi:10.1016/S0003-3472(05)80118-6

Campi KL, Greenberg GD, Kapoor A, Ziegler TE, Trainor BC. 2014. Sex differences in effects of dopamine D1 receptors on social withdrawal. *Neuropharmacology* **77:** 208–216. doi:10.1016/j.neuropharm.2013.09.026

Choleris E, Galea LAM, Sohrabji F, Frick KM. 2018. Sex differences in the brain: implications for behavioral and biomedical research. *Neurosci Biobehav Rev* **85:** 126–145. doi:10.1016/j.neubiorev.2017.07.005

Clark JW, Drummond SPA, Hoyer D, Jacobson LH. 2019. Sex differences in mouse models of fear inhibition: fear extinction, safety learning, and fear-safety discrimination. *Br J Pharmacol* **176:** 4149–4158. doi:10.1111/bph.14600

Coffey KR, Marx RG, Neumaier JF. 2019. DeepSqueak: a deep learning-based system for detection and analysis of ultrasonic vocalizations. *Neuropsychopharmacology* **44:** 859–868.

Darwin C. 1882. *The origin of species*. John Murray, London.

Datta SR, Anderson DJ, Branson K, Perona P, Leifer A. 2019. Computational neuroethology: a call to action. *Neuron* **104:** 11–24. doi:10.1016/j.neuron.2019.09.038

De Vries GJ, Boyle PA. 1998. Double duty for sex differences in the brain. *Behav Brain Res* **92:** 205–213.

Dolensek N, Gehrlach DA, Klein AS, Gogolla N. 2020. Facial expressions of emotion states and their neuronal correlates in mice. *Science* **368:** 89–94. doi:10.1126/science.aaz9468

Duque-Wilckens N, Steinman MQ, Laredo SA, Hao R, Perkeybile AM, Bales KL, Trainor BC. 2016. Inhibition of vasopressin V1a receptors in the medioventral bed nucleus of the stria terminalis has sex- and context-specific anxiogenic effects. *Neuropharmacology* **110:** 59–68. doi:10.1016/j.neuropharm.2016.07.018

Duque-Wilckens N, Steinman MQ, Busnelli M, Chini B, Yokoyama S, Pham M, Laredo SA, Hao R, Perkeybile AM, Minie VA, et al. 2018. Oxytocin receptors in the anteromedial bed nucleus of the stria terminalis promote stress-induced social avoidance in female California mice. *Biol Psychiatry* **83:** 203–213. doi:10.1016/j.biopsych.2017.08.024

Duque-Wilckens N, Torres LY, Yokoyama S, Minie VA, Tran AM, Petkova SP, Hao R, Ramos-Maciel S, Rios RA, Jackson K, et al. 2020. Extra-hypothalamic oxytocin neurons drive stress-induced social vigilance and avoidance. *Proc Natl Acad Sci* **117:** 26406–26413. doi:10.1073/pnas.2011890117

Emerson SB, Boyd SK. 1999. Mating vocalizations of female frogs: control and evolutionary mechanisms. *Brain Behav Evol* **53:** 187–197. doi:10.1159/000006594

Falkner AL, Dollar P, Perona P, Anderson DJ, Lin D. 2014. Decoding ventromedial hypothalamic neural activity during male mouse aggression. *J Neurosci* **34:** 5971–5984. doi:10.1523/JNEUROSCI.5109-13.2014

Falkner AL, Wei D, Song A, Watsek LW, Chen I, Chen P, Feng JE, Lin D. 2020. Hierarchical representations of aggression in a hypothalamic-midbrain circuit. *Neuron* **106:** 637–648.e6. doi:10.1016/j.neuron.2020.02.014

Forgie ML, Stewart J. 1993. Sex differences in amphetamine-induced locomotor activity in adult rats: Role of testosterone exposure in the neonatal period. *Pharmacol Biochem Behav* **46:** 637–645.

Friard O, Gamba M. 2016. BORIS: a free, versatile open-source event-logging software for video/audio coding and live observations. *Methods Ecol Evol* **7:** 1325–1330. doi:10.1111/2041-210X.12584

Goffinet J, Brudner S, Mooney R, Pearson J. 2021. Inferring low-dimensional latent descriptions of animal vocalizations. *eLife* **10:** e67855. doi:10.7554/eLife.67855

Golden SA, Covington HE III, Berton O, Russo SJ. 2011. A standardized protocol for repeated social defeat stress in mice. *Nat Protoc* **6:** 1183–1191. doi:10.1038/nprot.2011.361

Goodwill HL, Manzano-Nieves G, Gallo M, Lee HI, Oyerinde E, Serre T, Bath KG. 2019. Early life stress leads to sex differences in development of depressive-like outcomes in a mouse model. *Neuropsychopharmacology* **44:** 711–720. doi:10.1038/s41386-018-0195-5

Graving JM, Chae D, Naik H, Li L, Koger B, Costelloe BR, Couzin ID. 2019. DeepPoseKit, a software toolkit for fast and robust animal pose estimation using deep learning. *eLife* **8:** e47994. doi:10.7554/eLife.47994

Greenberg GD, Laman-Maharg A, Campi KL, Voigt H, Orr VN, Schaal L, Trainor BC. 2014. Sex differences in stress-induced social withdrawal: role of brain derived neurotrophic factor in the bed nucleus of the stria terminalis. *Front Behav Neurosci* **7:** 223. doi:10.3389/fnbeh.2013.00223

Grill HJ, Norgren R. 1978. The taste reactivity test. I: Mimetic responses to gustatory stimuli in neurologically normal rats. *Brain Res* **143:** 263–279. doi:10.1016/0006-8993(78)90568-1

Gruene TM, Flick K, Stefano A, Shea SD, Shansky RM. 2015. Sexually divergent expression of active and passive conditioned fear responses in rats. *eLife* **4:** e11352. doi:10.7554/eLife.11352

Hammerschmidt K, Radyushkin K, Ehrenreich H, Fischer J. 2012. The structure and usage of female and male mouse ultrasonic vocalizations reveal only minor differences. *PLoS ONE* **7:** e41133. doi:10.1371/journal.pone.0041133

Hashikawa K, Hashikawa Y, Tremblay R, Zhang J, Feng JE, Sabol A, Piper WT, Lee H, Rudy B, Lin D. 2017. Esr1[+] cells in the ventromedial hypothalamus control female aggression. *Nat Neurosci* **20:** 1580–1590. doi:10.1038/nn.4644

Henningsen JB, Poirel VJ, Mikkelsen JD, Tsutsui K, Simonneaux V, Gauer F. 2016. Sex differences in the photoperiodic regulation of RF-amide related peptide (RFRP) and its receptor GPR147 in the Syrian hamster. *J Comp Neurol* **524:** 1825–1838. doi:10.1002/cne.23924

Hodes GE, Pfau ML, Purushothaman I, Ahn HF, Golden SA, Christoffel DJ, Magida J, Brancato A, Takaashi A, Flanigan ME, et al. 2015. Sex differences in nucleus accumbens transcriptome profiles associated with susceptibility versus resilience to subchronic variable stress. *J Neurosci* **35:** 16362–16376. doi:10.1523/JNEUROSCI.1392-15.2015

Hoffman EJ, Turner KJ, Fernandez JM, Cifuentes D, Ghosh M, Ijaz S, Jain RA, Kubo F, Bill BR, Baier H, et al. 2016. Estrogens suppress a behavioral phenotype in zebrafish mutants of the autism risk gene, *CNTNAP2*. *Neuron* **89:** 725–733. doi:10.1016/j.neuron.2015.12.039

Holy TE, Guo Z. 2005. Ultrasonic songs of male mice. *PLoS Biol* **3:** e386. doi:10.1371/journal.pbio.0030386

Hsu AI, Yttri EA. 2021. An open source unsupervised algorithm for identification and fast prediction of behaviors. https://www.biorxiv.org/content/10.1101/770271v3 [accessed August 27, 2021].

Hubbard DT, Blanchard DC, Yang M, Markham CM, Gervacio A, Chun-I L, Blanchard RJ. 2004. Development of defensive behavior and conditioning to cat odor in the rat. *Physiol Behav* **80:** 525–530. doi:10.1016/j.physbeh.2003.10.006

Kelly KK. 1993. Androgenic induction of brain sexual dimorphism depends on photoperiod in meadow voles. *Physiol Behav* **53:** 245–249.

Khan MH, McDonagh J, Khan S, Shahabuddin M, Arora A, Khan FS, Shao L, Tzimiropoulos G. 2020. AnimalWeb: a large-scale hierarchical dataset of annotated animal faces. *Proceedings of the IEEE/CVF Conference on Computer Vision and Pattern Recognition (CVPR)*, pp. 6939–6948.

Klibaite U, Berman GJ, Cande J, Stern DL, Shaevitz JW. 2017. An unsupervised method for quantifying the behavior of paired animals. *Physical Biol* **14:** 015006. doi:10.1088/1478-3975/aa5c50

Krakauer JW, Ghazanfar AA, Gomez-Marin A, MacIver MA, Poeppel D. 2017. Neuroscience needs behavior: correcting a reductionist bias. *Neuron* **93:** 480–490. doi:10.1016/j.neuron.2016.12.041

Laman-Maharg A, Trainor BC. 2017. Stress, sex and motivated behaviors. *J Neurosci Res* **95:** 83–92. doi:10.1002/jnr.23815

Li S, Günel S, Ostrek M, Ramdya P, Fua P, Rhodin H. 2020. Deformation-aware unpaired image translation for pose estimation on laboratory animals. *Proceedings of the IEEE/CVF Conference on Computer Vision and Pattern Recognition (CVPR)*, pp. 13158–13168.

Lin D, Boyle MP, Dollar P, Lee H, Lein ES, Perona P, Anderson DJ. 2011. Functional identification of an aggression locus in the mouse hypothalamus. *Nature* **470:** 221–226. doi:10.1038/nature09736

Lischinsky JE, Lin D. 2020. Neural mechanisms of aggression across species. *Nat Neurosci* **23:** 1317–1328. doi:10.1038/s41593-020-00715-2

Mamlouk GM, Dorris DM, Barrett LR, Meitzen J. 2020. Sex bias and omission in neuroscience research is influenced by research model and journal, but not reported NIH funding. *Front Neuroendocrinol* **57:** 100835.

Maren S, De Oca B, Fanselow MS. 1994. Sex differences in hippocampal long-term potentiation (LTP) and Pavlovian fear conditioning in rats: positive correlation between LTP and contextual learning. *Brain Res* **661:** 25–34. doi:10.1016/0006-8993(94)91176-2

Martin P, Bateson P. 1994. *Measuring behavior, an introductory guide*, 2nd ed. Cambridge University Press, Cambridge.

Marr D, Nishihara HK. 1978. Representation and recognition of the spatial organization of three-dimensional shapes. *Proc R Soc Lond B Biol Sci* **200:** 269–294.

Mathis A, Mamidanna P, Cury KM, Abe T, Murthy VN, Weygandt Mathis M, Bethge M. 2018. DeepLabCut: markerless pose estimation of user-defined body parts with deep learning. *Nat Neurosci* **21:** 1281–1289. doi:10.1038/s41593-018-0209-y

Mathis A, Schneider S, Lauer J, Weygandt MM. 2020. A primer on motion capture with deep learning: principles, pitfalls, and perspectives. *Neuron* **108:** 44–65. doi:10.1016/j.neuron.2020.09.017

McCann KE, Huhman KL. 2012. The effect of escapable versus inescapable social defeat on conditioned defeat and social recognition in Syrian hamsters. *Physiol Behav* **105:** 493–497. doi:10.1016/j.physbeh.2011.09.009

Neunuebel JP, Taylor AL, Arthur BJ, Roian Egnor SE. 2015. Female mice ultrasonically interact with males during courtship displays. *eLife* **4:** e06203. doi:10.7554/eLife.06203

Newman EL, Covington HE, Suh J, Bicakci MB, Ressler KJ, DeBold JF, Miczek KA. 2019. Fighting females: neural and behavioral consequences of social defeat stress in female mice. *Biol Psychiatry* **86:** 657–668.

Nilsson SRO, Goodwin NL, Choong JJ, Hwang S, Wright HR, Norville ZC, Tong X, Lin D, Bentzley BS, Eshel N, et al. 2020. Simple behavioral analysis (SimBA)—an open source toolkit for computer classification of complex social behaviors in experimental animals. bioRxiv doi:10.1101/2020.04.19.049452

Odom KJ, Hall ML, Riebel K, Omland KE, Langmore NE. 2014. Female song is widespread and ancestral in songbirds. *Nat Commun* **5:** 3379. doi:10.1038/ncomms4379

Penna M, Capranica RR, Somers J. 1992. Hormone-induced vocal behavior and midbrain auditory sensitivity in the Green treefrog, *Hyla cinerea*. *J Comp Physiol A* **170:** 73–82. doi:10.1007/BF00190402

Pereira TD, Aldarondo DE, Willmore L, Kislin M, Wang SSH, Murthy M, Shaevitz JW. 2019. Fast animal pose estimation using deep neural networks. *Nat Methods* **16:** 117–125. doi:10.1038/s41592-018-0234-5

Pereira TD, Shaevitz JW, Murthy M. 2020. Quantifying behavior to understand the brain. *Nat Neurosci* **23:** 1537–1549. doi:10.1038/s41593-020-00734-z

Pernold K, Iannello F, Low BE, Rigamonti M, Rosati G, Scavizzi F, Wang J, Raspa M, Wiles MV, Ulfhake B. 2019. Towards large scale automated cage monitoring—diurnal rhythm and impact of interventions on in-cage activity of C57BL/6J mice recorded 24/7 with a non-disrupting capacitive-based technique. *PLoS ONE* **14:** e0211063. doi:10.1371/journal.pone.0211063

Perrot-Sinal TS, Kostenuik MA, Ossenkopp K-P, Kavaliers M. 1996. Sex differences in performance in the Morris water maze and the effects of initial nonstationary hidden platform training. *Behav Neurosci* 110: 1309–1320.

Prendergast BJ, Onishi KG, Zucker I. 2014. Female mice liberated for inclusion in neuroscience and biomedical research. *Neurosci Biobehav Rev* 40: 1–5. doi:10.1016/j .neubiorev.2014.01.001

Pryce CR, Lehmann J, Feldon J. 1999. Effect of sex on fear conditioning is similar for context and discrete CS in Wistar, Lewis and Fischer rat strains. *Pharmacol Biochem Behav* 64: 753–759. doi:10.1016/S0091-3057(99)00147-1

Roth ME, Carroll ME. 2004. Sex differences in the acquisition of IV methamphetamine self-administration and subsequent maintenance under a progressive ratio schedule in rats. *Psychopharmacology (Berl)* 172: 443–449. doi:10.1007/s00213-003-1670-0

Schulz KM, Sisk CL. 2016. The organizing actions of adolescent gonadal steroid hormones on brain and behavioral development. *Neurosci Biobehav Rev* 70: 148–158. doi:10 .1016/j.neubiorev.2016.07.036

Segalin C, Williams J, Karigo T, Hui M, Zelikowsky M, Sun JJ, Perona P, Anderson DJ, Kennedy A. 2020. The mouse action recognition system (MARS): a software pipeline for automated analysis of social behaviors in mice. bioRxiv doi:10.1101/2020.07.26.222299

Steinman MQ, Duque-Wilckens N, Greenberg GD, Hao R, Campi KL, Laredo SA, Laman-Maharg A, Manning CE, Doig IE, Lopez EM, et al. 2016. Sex-specific effects of stress on oxytocin neurons correspond with responses to intranasal oxytocin. *Biol Psychiatry* 80: 406–414. doi:10.1016/j.biopsych.2015.10.007

Steinman MQ, Duque-Wilckens N, Trainor BC. 2019. Complementary neural circuits for divergent effects of oxytocin: social approach versus social anxiety. *Biol Psychiatry* 85: 792–801. doi:10.1016/j.biopsych.2018.10.008

Sturman O, von Ziegler L, Schläppi C, Akyol F, Grewe B, Privitera M, Slominski D, Grimm C, Thieren L, Zerbi V, et al. 2020. Deep learning-based behavioral analysis reaches human accuracy and is capable of outperforming commercial solutions. *Neuropsychopharmacology* 45: 1942–1952. doi:10.1038/s41386-020-0776-y

Tobias ML, Marin ML, Kelley DB. 1993. The roles of sex, innervation, and androgen in laryngeal muscle of *Xenopus laevis*. *J Neurosci* 13: 324–333. doi:10.1523/JNEURO SCI.13-01-00324.1993

Trainor BC, Pride MC, Villalon Landeros R, Knoblauch NW, Takahashi EY, Silva AL, Crean KK. 2011. Sex differences in social interaction behavior following social defeat stress in the monogamous California mouse (*Peromyscus californicus*). *PLoS ONE* 6: e17405. doi:10.1371/journal .pone.0017405

Trainor BC, Takahashi EY, Campi KL, Florez SA, Greenberg GD, Laman-Maharg A, Laredo SA, Orr VN, Silva AL,

Steinman MQ. 2013. Sex differences in stress-induced social withdrawal: independence from adult gonadal hormones and inhibition of female phenotype by corncob bedding. *Horm Behav* 63: 543–550. doi:10.1016/j.yhbeh .2013.01.011

Van Segbroeck M, Knoll AT, Levitt P, Narayanan S. 2017. MUPET-mouse ultrasonic profile ExTraction: a signal processing tool for rapid and unsupervised analysis of ultrasonic vocalizations. *Neuron* 94: 465–485.e5. doi:10 .1016/j.neuron.2017.04.005

Van Swearingen AED, Walker QD, Kuhn CM. 2013. Sex differences in novelty- and psychostimulant-induced behaviors of C57BL/6 mice. *Psychopharmacology (Berl)* 225: 707–718. doi:10.1007/s00213-012-2860-4

Vogel AP, Tsanas A, Scattoni ML. 2019. Quantifying ultrasonic mouse vocalizations using acoustic analysis in a supervised statistical machine learning framework. *Sci Rep* 9: 8100. doi:10.1038/s41598-019-44221-3

von Ziegler L, Sturman O, Bohacek J. 2021. Big behavior: challenges and opportunities in a new era of deep behavior profiling. *Neuropsychopharmacology* 46: 33–44.

Warren BL, Mazei-Robison MS, Robison AJ, Iñiguez SD. 2020. Can I get a witness? Using vicarious defeat stress to study mood-related illnesses in traditionally understudied populations. *Biol Psychiatry* 88: 381–391. doi:10 .1016/j.biopsych.2020.02.004

Wilczynski W, Quispe M, Muñoz MI, Penna M. 2017. Arginine vasotocin, the social neuropeptide of amphibians and reptiles. *Front Endocrinol* 8: 186. doi:10.3389/fendo .2017.00186

Williams AV, Duque-Wilckens N, Ramos-Maciel S, Campi KL, Bhela SK, Xu CK, Jackson K, Chini B, Pesavento PA, Trainor BC. 2020. Social approach and social vigilance are differentially regulated by oxytocin receptors in the nucleus accumbens. *Neuropsychopharmacology* 45: 1423–1430. doi:10.1038/s41386-020-0657-4

Wiltschko AB, Tsukahara T, Zeine A, Anyoha R, Gillis WF, Markowitz JE, Peterson RE, Katon J, Johnson MJ, Datta SR. 2020. Revealing the structure of pharmacobehavioral space through motion sequencing. *Nat Neurosci* 23: 1433–1443.

Wright EC, Hostinar CE, Trainor BC. 2020. Anxious to see you: neuroendocrine mechanisms of social vigilance and anxiety during adolescence. *Eur J Neurosci* 52: 2516–2529. doi:10.1111/ejn.14628

Yohn CN, Dieterich A, Bazer AS, Maita I, Giedraitis M, Samuels BA. 2019. Chronic non-discriminatory social defeat is an effective chronic stress paradigm for both male and female mice. *Neuropsychopharmacology* 44: 2220–2229. doi:10.1038/s41386-019-0520-7

Epigenetic Mechanisms of Brain Sexual Differentiation

Bruno Gegenhuber[1,2] and Jessica Tollkuhn[1]

[1]Cold Spring Harbor Laboratory; [2]School of Biological Sciences, Cold Spring Harbor Laboratory, Cold Spring Harbor, New York 11724, USA

Correspondence: tollkuhn@cshl.edu

Across vertebrate species, gonadal hormones coordinate physiology with behavior to facilitate social interactions essential for reproduction and survival. In adulthood, these hormones activate neural circuits that regulate behaviors presenting differently in females and males, such as parenting and territorial aggression. Yet long before sex-typical behaviors emerge at puberty, transient hormone production during sensitive periods of neurodevelopment establish the circuits upon which adult hormones act. How transitory waves of early-life hormone signaling exert lasting effects on the brain remains a central question. Here we discuss how perinatal estradiol signaling organizes cellular and molecular sex differences in the rodent brain. We review classic anatomic studies revealing sex differences in cell number, volume, and neuronal projections, and consider how single-cell sequencing methods enable distinction between sex-biased cell-type abundance and gene expression. Finally, we highlight the recent discovery of a gene regulatory program activated by estrogen receptor α (ERα) following the perinatal hormone surge. A subset of this program displays sustained sex-biased gene expression and chromatin accessibility throughout the postnatal sensitive period, demonstrating a bona fide epigenetic mechanism. We propose that ERα-expressing neurons throughout the social behavior network use similar gene regulatory programs to coordinate brain sexual differentiation.

The gonadal steroid hormones testosterone, progesterone, and estrogens regulate reproductive maturation and are a principal source of sex-variable physiology and contributor to disease risk. In the brain, these hormones modulate the activity of neural circuits that control many behaviors, such as sexual and territorial behavior, care of offspring, food intake, and physical activity. Gonadal hormones also direct vertebrate brain sexual differentiation by acting during early-life critical periods to specify sex differences in brain development and neuroendocrine function (MacLusky and Naftolin 1981; Arnold and Breedlove 1985; Alward et al. 2018). Initial experiments demonstrated that exposure to testosterone during a perinatal window enhances the sensitivity of mating and aggression behaviors to adult testosterone. Fundamental behavioral responses to a social stimulus, such as male territorial aggression toward another male, are defined by the organizational effects of testosterone. These insights were followed by studies into hormonal regulation of sex differences in neuroanatomy; early life testosterone directs cell survival and death, neurite outgrowth, and synapse formation that emerge throughout a prolonged neurodevelopmental window. The ability of perinatal hormones to specify neuronal properties long after their clearance has led to the interpretation that brain sexual differentiation requires an epigenetic mechanism.

Many of the actions of steroid hormones occur through their cognate nuclear receptors, which principally act as transcription factors (TFs). These TFs are recruited to DNA, where they interact with additional cell-type-specific factors, as well as general transcription machinery, to activate or repress gene expression. Although genomic targets of hormone receptors have been well-characterized in specific cancers—androgen receptor (AR) in prostate cancer, ERα in breast, and endometrial cancer—they have been challenging to identify in the brain. In mice, estrogen receptor α (ERα) is the master regulator of brain sexual differentiation. Perinatal testosterone is converted to estradiol, the primary endogenous estrogen, in specific populations of neurons by the aromatase enzyme (Naftolin and Ryan 1975; Matsumoto et al. 2003). Female mice given estradiol at birth exhibit male levels of resident intruder aggression and urine-marking in adulthood (Wu et al. 2009), and male ERα mutants show deficits in these behaviors, as well as in sexual behavior (Ogawa et al. 1997; Wersinger et al. 1997; Wu and Tollkuhn 2017). In contrast, male mice do not express the receptor for testosterone, AR, in the brain during the perinatal hormone surge, and mice lacking AR in the nervous system still display masculine behaviors (Juntti et al. 2010). Interestingly, in rats, testosterone surges on embryonic day 18 (E18) (Weisz and Ward 1980), and AR is expressed in the social behavior network (SBN) by E20 (McAbee and DonCarlos 1998), suggesting the timing of neural aromatization and/or involvement of AR in masculinization (Dugger et al. 2007) differs across species.

How does perinatal estradiol set up sex differences in the brain? We propose two epigenetic perspectives to explain how perinatal estradiol influences brain development (Fig. 1). The original "epigenetic landscape" by C.H. Waddington depicts the developmental canalization of a cell through increasingly restricted states (Waddington 1940). By this original definition, "epigenetics" describes cellular memory—or self-perpetuation of a cellular phenotype following a transient signal (Henikoff and Greally 2016). In the context of brain sexual differentiation, neurons undergo permanent changes in their intrinsic properties as a consequence of transient estradiol production. For example, estradiol-exposed neurons may live instead of die, extend an axon to a distal target, or form synapses onto neighboring interneurons. All of these events represent components of brain sexual differentiation and can be studied with histological, circuit tracing, and high-resolution imaging approaches (Fig. 1, top). In the era of genome-wide sequencing, the definition of "epigenetics" has shifted to describe any persistent change in gene regulation following a transient signal. From this perspective, ERα recruitment to DNA in response to perinatal estradiol may alter chromatin modifications or structure at specific genomic regions, resulting in persistent sex-biased expression of effector genes responsible for differential neuronal phenotypes and/or responses to hormonal stimuli at puberty (Fig. 1, bottom). In this article, we discuss how both definitions of "epigenetics" apply to sexual differentiation of the brain. Neural sex differences are also influenced by sex chromosome complement, as discussed in Arnold (2022).

EPIGENETICS AS NEURONAL WIRING

Sex Differences in Neural Circuit Organization

There is sex variability in many aspects of behavior and physiology, but reproductive behaviors provide ideal phenotypes for functional dissection, as the neural substrates controlling these sex differences are well characterized. Several interconnected limbic brain regions, the SBN, integrate chemosensory information relayed from the vomeronasal organ with internal state and past experience to drive behavior. The SBN, which includes the posterodorsal medial amygdala (MeApd), principal nucleus of the bed

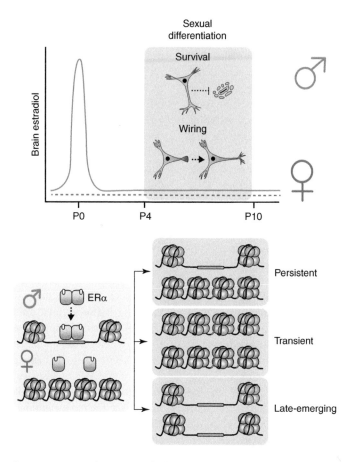

Figure 1. Cellular and epigenomic mechanisms underlying brain sexual differentiation. (*Top*) Schematic of brain estradiol levels in female (red) and male (blue) mice during the early postnatal critical period. Perinatal estradiol promotes brain sexual differentiation between P4 and P10 at the cellular level, including male-biased neuron survival and axon outgrowth/wiring. (*Bottom*) Perinatal estradiol promotes brain masculinization at the epigenomic level through activation of the transcription factor estrogen receptor α (ERα). Male-specific ERα activation leads to either sustained or transient sex differences in enhancer activity, resulting in sex-biased expression of genes involved in brain sexual differentiation.

nucleus of the stria terminalis (BNSTpr), medial preoptic area (MPOA), lateral septum, ventromedial hypothalamus (VMH), and midbrain nuclei, is largely conserved across vertebrate species (Newman 1999; Goodson 2005; O'Connell and Hofmann 2011). Combinatorial activation of neurons throughout this network by steroid hormones promotes seemingly disparate social behaviors, such as sexual receptivity and aggression.

The SBN contains the few brain regions that show reproducible, evolutionarily conserved anatomic sexual dimorphism, with quantifiable sex differences in cell number, cell/nucleus size, density, and/or projection strength or pattern. All of these regions express ERα, and most anatomic neural dimorphisms in rodents are a consequence of perinatal estradiol signaling. The MPOA is the most anterior portion of the hypothalamus and contains the first-identified and most-studied sexually dimorphic region: the sexually dimorphic nucleus (SDN-POA) (Raisman and Field 1971; Gorski et al. 1980). The SDN-POA has more cells and increased cell density in males compared to females, although the extent of this dimorphism varies across species (Campi et al. 2013). Similar to the SDN-POA, the BNSTpr is a key node in the SBN, linking MeApd, MPOA, and hypothalamus, and contains more neurons in males than in females (Hines et al. 1992; Zhou et al. 1995; Lebow and Chen 2016; Flanigan and Kash 2020). The male-biased cell number in these regions is a consequence of perinatal estradiol, which attenuates neuronal apoptosis in the first two postnatal weeks (Arai et al. 1996; Chung et al. 2000; Forger et al. 2004; Wu et al. 2009; Tsukahara et al. 2011). Treating female mice with estradiol on the day of birth increases cell number to a male-typical level, and males lacking aromatase have female-typical levels of apoptosis (Wu et al. 2009). Surprisingly, the effect of perinatal estradiol on cell survival signaling diverges across brain regions. For instance, the anteroventral periventricular (AVPV) nucleus of the hypothalamus, a region implicated in ovulation, is larger in females, due to a higher number of kisspeptin- and tyrosine hydroxylase (TH)-expressing neurons (Kauffman et al. 2007; Scott et al. 2015). Perinatal estradiol is responsible for female-biased cell number in the AVPV, as it induces caspase-mediated cell death in males (Waters and Simerly 2009). How the same signal exerts opposing effects across interconnected brain regions remains unknown.

Other regions within the SBN also contain structural dimorphisms, although some depend on adult hormones. For example, in rats, the increased size of the medial amygdala (MeA) in males is due to both organizational effects of estradiol and adult circulating testosterone (Johansen et al. 2004; Cooke 2006; Morris et al. 2008). Both soma size and dendrite length and/or branching may contribute to increased MeApd volume. Consistent with this observation, the MeApd of the seasonally breeding Siberian hamster also fluctuates in size concordant with seasonal changes in testosterone level (Cooke 2006). Sex differences in cell number within the rodent VMHvl have not been widely reported, but the anatomic organization of this region has been shown to differ between sexes, with females having a distinct neuronal composition within the lateral subdivision (Hashikawa et al. 2017).

In addition to regulating cell number and regional volume, perinatal estradiol regulates axon

growth and guidance, resulting in sex-biased neuronal connectivity throughout the SBN. One of the prime examples of this is the BNSTpr-to-AVPV projection, which in rats is stronger in males than in females, contrary to the higher cell number in the female AVPV (Hutton et al. 1998). This projection develops before postnatal day 10 (P10) (Hutton et al. 1998) and requires perinatal estradiol (Gu et al. 2003). Explant co-culture experiments further revealed that male, but not female, AVPV is sufficient to drive axon outgrowth from the BNSTpr of both sexes (Ibanez et al. 2001). The male BNSTpr also sends stronger projections to other regions in the SBN, including the ventral premammillary nucleus (PMV) and MeA, while innervation of the VMHvl and lateral septum appear similar between sexes (Gu et al. 2003). Differential BNSTpr targeting between downstream ERα-expressing regions suggests perinatal estradiol may induce region-specific axon guidance molecules to coordinate SBN wiring (Ibanez et al. 2001). Unlike the BNSTpr, the MPOA displays minimal sex differences in its outputs, as determined from classic fiber tract tracing or modern viral approaches (Simerly and Swanson 1988; Kohl et al. 2018), although projections specifically from the SDN-POA have not been investigated.

Neuron survival often depends on connectivity, in that integration into a functional network protects some neurons from programmed cell death. Mechanistically, this process has been shown to involve activity-dependent release of neurotrophic factors (e.g., BDNF) from target neurons, activity-dependent calcium influx of source neurons, or prosurvival signaling cascades downstream of axon guidance receptors (Raff et al. 1993; Vanderhaeghen and Cheng 2010; Priya et al. 2018). Likewise, during brain sexual differentiation, the connectivity of the BNSTpr-MeApd circuit may precede cell death across both regions. Cell death in female mice peaks at P4 in the MeApd and P7 in the BNSTpr (Wu et al. 2009), paralleling the ontogeny of connectivity between these regions in rats (Cooke and Simerly 2005). Cooke and Simerly detected MeApd projections to the BNSTpr ~1 wk before the reciprocal projections. MeApd axonal fibers were detected in the BNSTpr on E20, while reciprocal fibers from the BNSTpr did not appear until P5. The density of both connections increased rapidly during the postnatal period, such that adult levels of projections formed by P15. The specific molecular and/or activity-dependent mechanisms guiding this bidirectional connection remain unexplored.

Neuron Types: Dimorphism and Cell Identity

While studies over the past 40 years have discovered fundamental principles of brain sexual differentiation, many central concepts can now be revisited with advanced technology. Of note, progress in single-cell sequencing methodology, such as single-cell RNA-seq and ATAC-seq, has led to the discovery of tremendous diversity among neuron types both within and across brain regions. Multimodal studies, in which multiple properties of a single neuron or population of neurons are measured (e.g., morphology, electrophysiology, and transcriptome) indicate that transcriptionally defined neuron types display unique functional properties and/or developmental origins (Tasic et al. 2018; Hodge et al. 2019; Gouwens et al. 2020; BRAIN Initiative Cell Census Network (BICCN) 2021). Comparison of single-cell or single-nucleus RNA-seq data sets from the MPOA and BNST to established neuron types in the cortex and hippocampus has revealed the putative identity of hormone receptor–expressing neurons, providing initial clues as to their roles within the SBN. ERα-expressing neurons in the BNSTpr, MPOA, and MeApd are predominantly GABAergic (Wu and Tollkuhn 2017; Moffitt et al. 2018; Chen et al. 2019; Welch et al. 2019). In this article, we observe that BNST ERα-expressing neurons encompass populations transcriptomically resembling cortical interneuron types expressing *Vip* (Vip[+]), *Pvalb* (Pvalb[+]), or *Lamp5* (Lamp5[+]), as well as long-range projection neurons expressing *Sst* and *Chodl* (Sst[+]/Chodl[+]) (Gegenhuber et al. 2022). Among posterior BNST ERα[+] inhibitory neuron types, two, in particular, are more abundant in males than in females, marked by the expression of the TF nuclear factor I X-type (*Nfix*) or the secondary estradiol receptor, estrogen receptor β (*Esr2*/ERβ), respectively, consis-

tent with the known sexual dimorphism in the BNSTpr neuron number. ERα[+]/Nfix[+] neurons matched the identity of an analogous Lamp5[+] population in the SDN-POA, which is active in males during mating, aggression, and parenting (Moffitt et al. 2018), suggesting that sexual dimorphism in these regions is largely defined by a single Lamp5[+] interneuron type, otherwise known as a neurogliaform cell (Fig. 2, left panel; Tasic et al. 2018; Gegenhuber et al. 2022).

EPIGENETICS AS A PERSISTENT GENE REGULATORY PROGRAM

Brain sexual differentiation occurs over a timescale of weeks, while perinatal hormone subsides within hours. What enables a neuron to maintain a sex bias in its intrinsic properties long after receiving a transient signal? Our second perspective on epigenetic regulation of brain sexual differentiation proposes that ERα binding to DNA in response to perinatal estradiol induces an "epigenomic memory," or a persistent gene regulatory program that enables male-typical neuronal wiring throughout the early postnatal critical period.

Epigenomic mechanisms, such as regulation of histone modifications or DNA methylation, have proven challenging to study in the brain, due to the heterogeneity of neuron types and states in bulk tissue preparations. While prior work suggests brain sexual differentiation involves epigenomic regulation (McCarthy et al. 2009; McCarthy and Nugent 2015), details of this process, such as the identity of enhancers and transcriptional complexes activated by perinatal estradiol, as well as neuron types involved, remain elusive (Gegenhuber and Tollkuhn 2019).

Two recent studies provide convincing evidence of long-lasting changes in neuronal gene expression in response to a transient stimulus. Both studies employed inducible, activity-dependent genetic labeling approaches to capture and sequence engram cells involved in fear conditioning (Chen et al. 2020; Marco et al. 2020). In the first study, the Quake and Südhof laboratories trained mice in a conditioning chamber, and then labeled neurons that were activated during fear memory recall 16 d later. Following an ad-

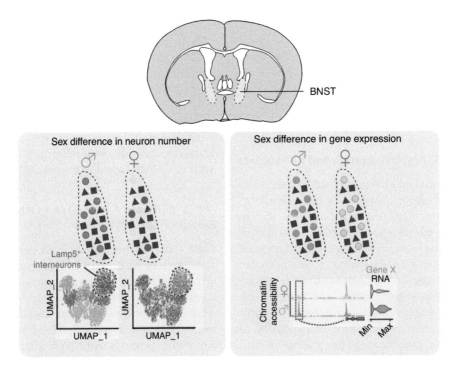

Figure 2. Profiling the principal nucleus of the bed nucleus of the stria terminalis (BNSTpr) sexual differentiation at single-cell resolution. Single-cell sequencing methods reveal two mechanisms underlying sexual differentiation of the BNSTpr. (*Left*) Single-nucleus RNA-sequencing (snRNA-seq) reveals a male-bias in the abundance of a specific neuron type (indicated in purple) within the BNSTpr, which is visualized in a two-dimensional space (represented by the unitless UMAP-1 and UMAP-2 axes) using the Uniform Manifold Approximation and Projection (UMAP) method. UMAP first constructs a high-dimensional weighted graph representation of the snRNA-seq data, in which the edge connections between each cell are determined by local radii that are proportional to the distance to each cell's *k*th nearest neighbor. This high-dimensional representation is subsequently projected onto low-dimensional space, preserving cell cluster information by prioritizing high-probability edge weights. (*Right*) Single-nucleus multiome (ATAC + RNA) sequencing enables simultaneous detection of sex differences in gene expression and enhancer activity within specific neuron types. An enhancer and gene with male-biased accessibility and expression, respectively, are depicted in the browser track *below*.

ditional 9 d to permit cortical consolidation, labeled neurons in the anterior cingulate cortex (ACC) of the medial prefrontal cortex (mPFC) were captured for single-cell RNA-seq. Relative to home cage, non-fear-conditioned, and non-recall control groups, fear memory recall–activated neurons increased expression of a particular subset of vesicle exocytosis and neuronal communication genes, suggesting remote memory activates a persistent transcriptional program. Surprisingly, nonneuronal cells, including astrocytes and microglia, displayed persistent elevated expression of lipid metabolism, glucose transport, and cytoskeletal organization genes, indicating communication between nonneuronal cells and

neurons may support long-term memory. In the second study, the Tsai laboratory used a similar strategy, labeling hippocampal neurons that activated during training, memory consolidation, or memory recall. By collecting epigenomic and chromatin architecture data, in addition to gene expression, they observed that fear conditioning generates persistent increases in chromatin accessibility and enhancer–promoter contacts, which primes neurons to rapidly transcribe genes in response to engram reactivation.

These studies establish the fundamental principle that neurons can encode transient stimuli within the epigenome to influence future responses. This principle can now be examined in the context of a different transient stimulus: developmental hormone signaling. In this section, we discuss prior insights into sex-biased and/or hormone-regulated genes in the brain, and review a recent study using an array of TF profiling and single-cell sequencing approaches to identify an epigenetic mechanism of brain sexual differentiation.

Sex Differences in Gene Expression in the Brain

Several genes display reproducible sex differences in expression within the SBN, including all four gonadal hormone nuclear receptors and several neuropeptides (for review, see Gegenhuber and Tollkuhn 2020). Additional sex-biased genes have recently been discovered using single-cell or single-nucleus RNA-seq. To achieve the statistical power necessary to detect sex-biased expression, studies have either purified and sequenced a select number of genetically labeled cell populations (Chen et al. 2019; Kim et al. 2019; van Veen et al. 2020), or sequenced a large number of cells (>100,000) from wild-type animals (Welch et al. 2019). Notably, sex differences in gene expression have not yet been described in the extensive cortical and hippocampal data sets generated by the Allen Institute (Yao et al. 2021); although some sex differences in gene isoforms, which can only be detected with full-length sequencing methods such as SMART-seq, have been observed (Booeshaghi et al. 2021). In addition to discovering the identity of male-biased neuron types in the BNSTpr, single-nucleus RNA-seq revealed extensive and heterogeneous sex differences in gene expression across inhibitory neuron types (Welch et al. 2019; Gegenhuber et al. 2022). These genes associate with sex-biased putative enhancers that require male gonadal hormones (Gegenhuber et al. 2022). Collectively, single-cell studies demonstrate that sex differences in gene expression occur exclusively in cell types expressing hormone receptors (Kim et al. 2019; van Veen et al. 2020; Gegenhuber et al. 2022). In the BNSTpr, *Esr2*, *Pgr*, and *Ar* are all sex-biased and expressed within ERα neurons, suggesting the primacy of ERα in orchestrating sex differences in the mouse brain.

A functional link between developmentally programmed sex differences and adult physiology has recently been described in the VMHvl (van Veen et al. 2020). In addition to regulating social behaviors, VMHvl ERα-expressing neurons regulate thermogenesis and physical activity across the estrous cycle (Correa et al. 2015). Single-cell RNA-seq profiling of the VMHvl revealed two female-biased neuron types, marked by *Rprm* and *Tac1*, respectively. A separate study also replicated the presence of sex-biased neuron types in the VMHvl, using a combination of high-throughput and high-depth single-cell RNA-seq approaches (Kim et al. 2019). Using the four-core genotypes model, which distinguishes between the effects of sex chromosomes and gonadal sex, van Veen et al. further demonstrated that gonadal hormones act organizationally to cause female-biased expression of *Rprm* and *Tac1*. ERα deletion in the male VMHvl also feminized expression of these two genes, suggesting that perinatal ERα signaling either permanently represses their expression or promotes apoptosis of neurons expressing these markers. Future studies incorporating hormone receptor binding and chromatin data may reveal whether these findings indeed reflect sex differences in neuron identity or rather sex-differential expression within shared neuron types. Remarkably, van Veen et al. found that *Rprm* knockdown in the VMHvl increased core body temperature in females but not in males, while ERα mutant males, which had elevated *Rprm* expression, decreased core body temperature. These findings suggest that perinatal ERα activation may permanently alter the expression of genes or abundance of neuron types that contribute to distinct sex-differential behaviors.

Neural Targets of ERα

What are the genomic targets of ERα in the brain? This question, while appearing simple, has proven elusive, due to the high-input requirement for profiling TF binding with chromatin immunoprecipitation (ChIP). For this reason, prior ERα ChIP-seq studies have been restricted to cancer cell lines and nonneural tissues with abundant ERα expression (Hurtado et al. 2011; Gertz et al. 2013). Collectively, these studies revealed ERα, and other hormone receptors, regulate cell-type-specific gene expression programs by binding intergenic and intronic enhancers, rather than promoters, often in collaboration with cell-type-specific TF complexes. Due to the lack of promoter binding, hormone receptor–binding patterns must be identified in the brain to distinguish direct from indirect effects of hormone signaling on neural gene regulation, neurodevelopment, and behavior.

A newer, low-input chromatin immuno-cleavage-based approach for genome-wide profiling of TF binding, cleavage under targets and release using nuclease (CUT&RUN), has recently enabled detection of ERα genomic binding sites in the brain for the first time (Skene and Henikoff 2017; Gegenhuber et al. 2022). CUT&RUN uses fresh, unfixed cells or nuclei immobilized on magnetic beads and relies on a protein A-micrococcal nuclease fusion protein to cleave DNA surrounding an antibody-bound TF of interest in the presence of calcium. Cleaved DNA fragments diffuse out of the nucleus into the supernatant, leaving the undigested chromatin and associated background signal inside the bead-bound nuclei. The lack of fixation, sonication, and immunoprecipitation steps in CUT&RUN leads to increased specificity and sensitivity compared to ChIP-seq, thereby enabling TF profiling from fewer cells. The authors used CUT&RUN to detect ERa genomic binding in gonadectomized adult mice treated acutely with estradiol or vehicle control.

ERα predominantly bound sites unique to the brain: ~60% were not detected in other ERα-expressing tissues such as mammary gland, uterus, or liver. These neural-specific ERα target genes include ion channels, neurotransmitter receptors, cell adhesion molecules, and other genes associated with neuroplasticity and synaptic organization. In the future, exploration of ERα target genes in different brain regions and physiological states may link their function to estradiol-regulated phenotypes. For example, up-regulation of a single ERα target gene, melanocortin 4 receptor (Mc4r), in the VMHvl increases spontaneous locomotion (Krause et al. 2021). By targeting a guide RNA directly to the

estrogen response element in the promoter of Mc4r, Krause et al. used CRISPR activation (CRISPRa) to elevate Mc4r expression specifically within the VMHvl, resulting in increased physical activity in both sexes.

Unexpectedly, we found that in the absence of adult gonadal hormones, females and males display a similar genomic response to estradiol. Adult, gonadectomized mice receiving an acute estradiol injection activated similar genomic responses across sexes; estradiol induced ERα binding and chromatin accessibility at the same loci in the BNSTpr of females and males. These data suggest that while early-life hormones permanently organize certain features of SBN circuit wiring, they do not restrict the potential to respond to alterations in the hormonal milieu in adult life. Consistent with this, gonadectomized female rats and mice administered testosterone or estradiol in adulthood display male-typical mounting behaviors; in mice, this phenotype requires ERα (Södersten 1972; Wersinger et al. 1997).

A Gene Regulatory Signature of Perinatal Estradiol

Despite sex-shared genomic responses to exogenous estradiol in adulthood, perinatal hormones establish thresholds for the display of reproductive behaviors, presumably through their lasting effects on the connectivity and number of SBN neurons. Which genes are regulated by perinatal estradiol, and can they persist after this transient signal subsides? Using a genetic labeling approach to selectively capture and profile BNST ERα-expressing neurons in mice, the authors examined gene expression and chromatin accessibility on P4—4 d after perinatal hormone becomes undetectable in serum—in females and males treated with vehicle control at birth, and females treated with estradiol. At this time, both gene expression and accessibility of putative enhancers differ substantially between sexes, and treating females with estradiol largely masculinized these features. Furthermore, by profiling ERα binding on P0, the authors found that perinatal estradiol induces ERα recruitment to male-biased enhancers, demonstrating that the brief pulse of testosterone at birth manifests in the

BNST as a subsequent wave of enhancer activation, resulting in sex-biased gene expression (Gegenhuber et al. 2022).

As mentioned previously, BNST ERα-expressing neurons constitute multiple, discrete neuron types, while experiments described thus far were performed on the bulk population. Does ERα control distinct gene regulatory programs across these neuron types, thereby diversifying cellular responses to perinatal estradiol? Using single-cell sequencing approaches, as schematized in Figure 2, the authors discovered striking heterogeneity in sex-biased genes, and their corresponding enhancers across neuron types, demonstrating ERα binds different genomic regions across these types in response to the same hormone. Performing single-cell experiments again on P14 further revealed that a subset of sex-biased, ERα target genes persists as sex-biased throughout the neonatal sensitive period. These include genes involved in axon outgrowth (*Crim1*, *Nell2*, *Pak7*/PAK5), synapse formation (*Col25a1*, *Il1rap*), synaptic transmission (*Asic2*, *Kcnab1*, *Scg2*, *Shisa6*), and transcriptional regulation (*Greb1*, *Plagl1*, *Sox5*). Selective deletion of ERα in inhibitory neurons both feminized the expression of these genes and the abundance of male-biased neuron types. Together, this study reveals two epigenetic mechanisms by which ERα controls brain sexual differentiation: regulating the abundance of two BNSTpr inhibitory neuron types, and organizing a persistent gene regulatory program within these types (Gegenhuber et al. 2022).

Puberty represents an additional critical period for brain sexual differentiation (Schulz and Sisk 2016; Piekarski et al. 2017). We speculate that the sharp rise in gonadal hormones during this time may reorganize gene regulatory programs, if not anatomy and/or synapse formation, to facilitate sex-differential circuit function and behavior in adulthood.

CONCLUDING REMARKS: A SPECTRUM OF SEX DIFFERENCES

Here we reviewed the cellular and genomic events set in motion by developmental hormone signaling. We propose that brain sexual differentiation constitutes an epigenetic process, according to both traditional and contemporary definitions of the term. Going forward, the ultimate goal will be to harmonize these definitions by determining how persistent, sex-biased hormone receptor target genes control specific cellular and/or behavioral sex differences. Thus far, the gene regulatory program activated by ERα has been uncovered in the developing BNST. Future exploration will undoubtedly reveal how regulation of gene expression in other regions and neuron types causes brain sexual differentiation. Indeed, organizational effects of gonadal hormones have been described for brain regions outside the SBN, such as regulation of *Pomc* and *Kiss1* expression in the arcuate nucleus (Nohara et al. 2011), ACh release in the hippocampus (Mitsushima et al. 2009), and postnatal hippocampal neurogenesis (Bowers et al. 2010). In the cortex, ERα and ERβ (Gerlach et al. 1983; Shughrue et al. 1990; Miranda and Toran-Allerand 1992; Yokosuka et al. 1995; Pérez et al. 2003), AR (Nuñez et al. 2003), and progesterone receptor (PR) (López and Wagner 2009) show dynamic, layer-specific expression, but certain receptors (e.g., ERα) subside prior to weaning; potentially, each orchestrates unique organizational programs across cortical areas.

Transcriptomic and epigenomic characterization of BNST ERα-expressing neurons revealed that sexual differentiation does not constrain the capacity to mount a genomic response to estradiol in adulthood. Likewise, manipulation of hormone levels, chemosensory input, or neuronal activity causes both sexes to engage in behaviors typical of the opposite sex, once considered "sex-specific" (Wei et al. 2021; Zilkha et al. 2021). These findings converge upon a model of behavioral sex differences as probabilistic rather than absolute: perinatal estradiol establishes differential gating or thresholding of shared neural circuitry, providing quantitative variation in behavioral intensity as opposed to qualitative presence or absence of hard-wired behaviors.

As the field continues studying the SBN of other species, it will be worth revisiting long-standing assumptions about the relationship between neural and behavioral sex differences. For example, is a male bias in SDN-POA anatomy required for male-typical sexual behavior? While the SDN-POA was initially implicated in male sexual behavior, female mice supplemented with testosterone or estradiol in adulthood show male-typical levels of mounting toward other females (Baum 2009), as do females that are unable to detect pheromonal cues (Kimchi et al. 2007). In addition, rodent species that display more affiliative behaviors, such as biparental care and pair-bonding, have reduced SDN-POA dimorphism compared to mice and rats, yet still engage in territorial aggression and sex-typical mating behaviors. In particular, prairie voles lack sexual dimorphism in the SDN-POA (Shapiro et al. 1991), whereas California mice (*Peromyscus californicus*) have intermediate levels of dimorphism in this region (Campi et al. 2013). Males of these two species also produce low levels of circulating testosterone (<1 ng/mL); however, both sexes can display territorial aggression, which in mice and rats is developmentally programmed by perinatal estradiol and activated by testosterone in adulthood. Finally, the timing of the perinatal testosterone surge differs between mice and rats, which likely contributes to differences in the display of SBN-mediated behaviors. We hypothesize that behavioral variation arises, in part, as a consequence of distinct gene regulatory programs invoked by steroid hormones in the developing and adult brains of different species.

Remarkably, one of the earliest writings on "epigenetics" described sexual differentiation. In *Organisers and Genes*, C.H. Waddington (1940) used the example of size differences in dog breeds to discuss the genetic basis of biological variability. With no knowledge of the genomic actions of hormones, or of the nature of genes, he postulated that variation in dog sizes depends on genetically controlled variation in hormone production from the pituitary. He further speculated that "the intersexual condition is no more a case of sharp alternatives than is pituitary dwarfism. It becomes merely one of the numerous cases in which the normal well-defined alternatives are disrupted by changes in the genotypic system on which they are based." These "sharp alternatives" were depicted as valleys of canalization in the classic illustration of the epigenetic landscape, which Waddington describes immediately following his musings on sexual differentiation. Given the extensive differences in social behavior across and within vertebrate species, we propose that a corresponding abundance of epigenetic programs generates diversity in sexually differentiated end points and adult states.

REFERENCES

*Reference is also in this subject collection.

Alward BA, Cornil CA, Balthazart J, Ball GF. 2018. The regulation of birdsong by testosterone: multiple timescales and multiple sites of action. *Horm Behav* **104:** 32–40. doi:10.1016/j.yhbeh.2018.04.010

Arai Y, Sekine Y, Murakami S. 1996. Estrogen and apoptosis in the developing sexually dimorphic preoptic area in female rats. *Neurosci Res* **25:** 403–407. doi:10.1016/0168-0102(96)01070-X

* Arnold AP. 2022. Integrating sex chromosome and endocrine theories to improve teaching of sexual differentiation. *Cold Spring Harb Perspect Biol* doi:10.1101/cshperspect.a039057

Arnold AP, Breedlove SM. 1985. Organizational and activational effects of sex steroids on brain and behavior: a reanalysis. *Horm Behav* **19:** 469–498. doi:10.1016/0018-506X(85)90042-X

Baum MJ. 2009. Sexual differentiation of pheromone processing: links to male-typical mating behavior and partner preference. *Horm Behav* **55:** 579–588. doi:10.1016/j.yhbeh.2009.02.008

Booeshaghi AS, Yao Z, van Velthoven C, Smith K, Tasic B, Zeng H, Pachter L. 2021. Isoform cell-type specificity in the mouse primary motor cortex. *Nature* **598:** 195–199. doi:10.1038/s41586-021-03969-3

Bowers JM, Waddell J, McCarthy MM. 2010. A developmental sex difference in hippocampal neurogenesis is mediated by endogenous oestradiol. *Biol Sex Differ* **1:** 8. doi:10.1186/2042-6410-1-8

BRAIN Initiative Cell Census Network (BICCN). 2021. A multimodal cell census and atlas of the mammalian primary motor cortex. *Nature* **598:** 86–102. doi:10.1038/s41586-021-03950-0

Campi KL, Jameson CE, Trainor BC. 2013. Sexual dimorphism in the brain of the monogamous California mouse (*Peromyscus californicus*). *Brain Behav Evol* **81:** 236–249. doi:10.1159/000353260

Chen PB, Hu RK, Wu YE, Pan L, Huang S, Micevych PE, Hong W. 2019. Sexually dimorphic control of parenting behavior by the medial amygdala. *Cell* **176:** 1206–1221. e18. doi:10.1016/j.cell.2019.01.024

Chen MB, Jiang X, Quake SR, Südhof TC. 2020. Persistent transcriptional programmes are associated with remote memory. *Nature* **587:** 437–442. doi:10.1038/s41586-020-2905-5

Chung WC, Swaab DF, De Vries GJ. 2000. Apoptosis during sexual differentiation of the bed nucleus of the stria terminalis in the rat brain. *J Neurobiol* **43:** 234–243. doi:10

Cite this article as *Cold Spring Harb Perspect Biol* doi: 10.1101/cshperspect.a039099

.1002/(SICI)1097-4695(20000605)43:3<234::AID-NEU2
>3.0.CO;2-3

Cooke BM. 2006. Steroid-dependent plasticity in the medial amygdala. *Neuroscience* **138:** 997–1005. doi:10.1016/j .neuroscience.2005.06.018

Cooke BM, Simerly RB. 2005. Ontogeny of bidirectional connections between the medial nucleus of the amygdala and the principal bed nucleus of the stria terminalis in the rat. *J Comp Neurol* **489:** 42–58. doi:10.1002/cne.20612

Correa SM, Newstrom DW, Warne JP, Flandin P, Cheung CC, Lin-Moore AT, Pierce AA, Xu AW, Rubenstein JL, Ingraham HA. 2015. An estrogen-responsive module in the ventromedial hypothalamus selectively drives sex-specific activity in females. *Cell Rep* **10:** 62–74. doi:10 .1016/j.celrep.2014.12.011

Dugger BN, Morris JA, Jordan CL, Breedlove SM. 2007. Androgen receptors are required for full masculinization of the ventromedial hypothalamus (VMH) in rats. *Horm Behav* **51:** 195–201. doi:10.1016/j.yhbeh.2006.10.001

Flanigan ME, Kash TL. 2020. Coordination of social behaviors by the bed nucleus of the stria terminalis. *Eur J Neurosci* doi: 10.1111/ejn.14991

Forger NG, Rosen GJ, Waters EM, Jacob D, Simerly RB, de Vries GJ. 2004. Deletion of *Bax* eliminates sex differences in the mouse forebrain. *Proc Natl Acad Sci* **101:** 13666–13671. doi:10.1073/pnas.0404644101

Gegenhuber B, Tollkuhn J. 2019. Sex differences in the epigenome: a cause or consequence of sexual differentiation of the brain? *Genes (Basel)* **10:** 432. doi:10.3390/genes 10060432

Gegenhuber B, Tollkuhn J. 2020. Signatures of sex: sex differences in gene expression in the vertebrate brain. *Wiley Interdiscip Rev Dev Biol* **9:** e348. doi:10.1002/wdev.348

Gegenhuber B, Wu MV, Bronstein R, Tollkuhn J. 2022. Gene regulation by gonadal hormone receptors underlies brain sex differences. *Nature* doi:10.1038/s41586-022-04686-1

Gerlach JL, McEwen BS, Toran-Allerand CD, Friedman WJ. 1983. Perinatal development of estrogen receptors in mouse brain assessed by radioautography, nuclear isolation and receptor assay. *Brain Res* **11:** 7–18. doi:10.1016/ 0165-3806(83)90197-9

Gertz J, Savic D, Varley KE, Partridge EC, Safi A, Jain P, Cooper GM, Reddy TE, Crawford GE, Myers RM. 2013. Distinct properties of cell-type-specific and shared transcription factor binding sites. *Mol Cell* **52:** 25–36. doi:10 .1016/j.molcel.2013.08.037

Goodson JL. 2005. The vertebrate social behavior network: evolutionary themes and variations. *Horm Behav* **48:** 11–22. doi:10.1016/j.yhbeh.2005.02.003

Gorski RA, Harlan RE, Jacobson CD, Shryne JE, Southam AM. 1980. Evidence for the existence of a sexually dimorphic nucleus in the preoptic area of the rat. *J Comp Neurol* **193:** 529–539. doi:10.1002/cne.901930214

Gouwens NW, Sorensen SA, Baftizadeh F, Budzillo A, Lee BR, Jarsky T, Alfiler L, Baker K, Barkan E, Berry K, et al. 2020. Integrated morphoelectric and transcriptomic classification of cortical GABAergic cells. *Cell* **183:** 935–953. e19. doi:10.1016/j.cell.2020.09.057

Gu G, Cornea A, Simerly RB. 2003. Sexual differentiation of projections from the principal nucleus of the bed nuclei of

the stria terminalis. *J Comp Neurol* **460:** 542–562. doi:10 .1002/cne.10677

Hashikawa K, Hashikawa Y, Tremblay R, Zhang J, Feng JE, Sabol A, Piper WT, Lee H, Rudy B, Lin D. 2017. Esr1[+] cells in the ventromedial hypothalamus control female aggression. *Nat Neurosci* **20:** 1580–1590. doi:10.1038/nn.4644

Henikoff S, Greally JM. 2016. Epigenetics, cellular memory and gene regulation. *Curr Biol* **26:** R644–R648. doi:10 .1016/j.cub.2016.06.011

Hines M, Allen LS, Gorski RA. 1992. Sex differences in subregions of the medial nucleus of the amygdala and the bed nucleus of the stria terminalis of the rat. *Brain Res* **579:** 321–326. doi:10.1016/0006-8993(92)90068-K

Hodge RD, Bakken TE, Miller JA, Smith KA, Barkan ER, Graybuck LT, Close JL, Long B, Johansen N, Penn O, et al. 2019. Conserved cell types with divergent features in human versus mouse cortex. *Nature* **573:** 61–68. doi:10 .1038/s41586-019-1506-7

Hurtado A, Holmes KA, Ross-Innes CS, Schmidt D, Carroll JS. 2011. FOXA1 is a key determinant of estrogen receptor function and endocrine response. *Nat Genet* **43:** 27–33. doi:10.1038/ng.730

Hutton LA, Gu G, Simerly RB. 1998. Development of a sexually dimorphic projection from the bed nuclei of the stria terminalis to the anteroventral periventricular nucleus in the rat. *J Neurosci* **18:** 3003–3013. doi:10 .1523/JNEUROSCI.18-08-03003.1998

Ibanez MA, Gu G, Simerly RB. 2001. Target-dependent sexual differentiation of a limbic-hypothalamic neural pathway. *J Neurosci* **21:** 5652–5659. doi:10.1523/JNEUROSCI .21-15-05652.2001

Johansen JA, Jordan CL, Breedlove SM. 2004. Steroid hormone masculinization of neural structure in rats: a tale of two nuclei. *Physiol Behav* **83:** 271–277. doi:10.1016/j .physbeh.2004.08.016

Juntti SA, Tollkuhn J, Wu MV, Fraser EJ, Soderborg T, Tan S, Honda S-I, Harada N, Shah NM. 2010. The androgen receptor governs the execution, but not programming, of male sexual and territorial behaviors. *Neuron* **66:** 260–272. doi:10.1016/j.neuron.2010.03.024

Kauffman AS, Gottsch ML, Roa J, Byquist AC, Crown A, Clifton DK, Hoffman GE, Steiner RA, Tena-Sempere M. 2007. Sexual differentiation of Kiss1 gene expression in the brain of the rat. *Endocrinology* **148:** 1774–1783. doi:10.1210/en.2006-1540

Kim D-W, Yao Z, Graybuck LT, Kim TK, Nguyen TN, Smith KA, Fong O, Yi L, Koulena N, Pierson N, et al. 2019. Multimodal analysis of cell types in a hypothalamic node controlling social behavior. *Cell* **179:** 713–728.e17. doi:10.1016/j.cell.2019.09.020

Kimchi T, Xu J, Dulac C. 2007. A functional circuit underlying male sexual behaviour in the female mouse brain. *Nature* **448:** 1009–1014. doi:10.1038/nature06089

Kohl J, Babayan BM, Rubinstein ND, Autry AE, Marin-Rodriguez B, Kapoor V, Miyamishi K, Zweifel LS, Luo L, Uchida N, et al. 2018. Functional circuit architecture underlying parental behaviour. *Nature* **556:** 326–331. doi:10.1038/s41586-018-0027-0

Krause WC, Rodriguez R, Gegenhuber B, Matharu N, Rodriguez AN, Padilla-Roger AM, Toma K, Herber CB, Correa SM, Duan X, et al. 2021. Oestrogen engages brain

MC4R signalling to drive physical activity in female mice. *Nature* **599**: 131–135. doi:10.1038/s41586-021-04010-3

Lebow MA, Chen A. 2016. Overshadowed by the amygdala: the bed nucleus of the stria terminalis emerges as key to psychiatric disorders. *Mol Psychiatry* **21**: 450–463. doi:10.1038/mp.2016.1

López V, Wagner CK. 2009. Progestin receptor is transiently expressed perinatally in neurons of the rat isocortex. *J Comp Neurol* **512**: 124–139. doi:10.1002/cne.21883

MacLusky NJ, Naftolin F. 1981. Sexual differentiation of the central nervous system. *Science* **211**: 1294–1302. doi:10.1126/science.6163211

Marco A, Meharena HS, Dileep V, Raju RM, Davila-Velderrain J, Zhang AL, Adaikkan C, Young JZ, Gao F, Kellis M, et al. 2020. Mapping the epigenomic and transcriptomic interplay during memory formation and recall in the hippocampal engram ensemble. *Nat Neurosci* **23**: 1606–1617. doi:10.1038/s41593-020-00717-0

Matsumoto T, Honda SI, Harada N. 2003. Alteration in sex-specific behaviors in male mice lacking the aromatase gene. *Neuroendocrinology* **77**: 416–424. doi:10.1159/000071313

McAbee MD, DonCarlos LL. 1998. Ontogeny of region-specific sex differences in androgen receptor messenger ribonucleic acid expression in the rat forebrain. *Endocrinology* **139**: 1738–1745. doi:10.1210/endo.139.4.5940

McCarthy MM, Nugent BM. 2015. At the frontier of epigenetics of brain sex differences. *Front Behav Neurosci* **9**: 221. doi:10.3389/fnbeh.2015.00221

McCarthy MM, Auger AP, Bale TL, De Vries GJ, Dunn GA, Forger NG, Murray EK, Nugent BM, Schwarz JM, Wilson ME. 2009. The epigenetics of sex differences in the brain. *J Neurosci* **29**: 12815–12823. doi:10.1523/JNEUROSCI.3331-09.2009

Miranda RC, Toran-Allerand CD. 1992. Developmental expression of estrogen receptor mRNA in the rat cerebral cortex: a nonisotopic in situ hybridization histochemistry study. *Cereb Cortex* **2**: 1–15. doi:10.1093/cercor/2.1.1

Mitsushima D, Takase K, Funabashi T, Kimura F. 2009. Gonadal steroids maintain 24 h acetylcholine release in the hippocampus: organizational and activational effects in behaving rats. *J Neurosci* **29**: 3808–3815. doi:10.1523/JNEUROSCI.5301-08.2009

Moffitt JR, Bambah-Mukku D, Eichhorn SW, Vaughn E, Shekhar K, Perez JD, Rubinstein ND, Hao J, Regev A, Dulac C, et al. 2018. Molecular, spatial, and functional single-cell profiling of the hypothalamic preoptic region. *Science* **362**: eaau5324. doi:10.1126/science.aau5324

Morris JA, Jordan CL, Breedlove SM. 2008. Sexual dimorphism in neuronal number of the posterodorsal medial amygdala is independent of circulating androgens and regional volume in adult rats. *J Comp Neurol* **506**: 851–859. doi:10.1002/cne.21536

Naftolin F, Ryan KJ. 1975. The metabolism of androgens in central neuroendocrine tissues. *J Steroid Biochem* **6**: 993–997. doi:10.1016/0022-4731(75)90340-4

Newman SW. 1999. The medial extended amygdala in male reproductive behavior. A node in the mammalian social behavior network. *Ann NY Acad Sci* **877**: 242–257. doi:10.1111/j.1749-6632.1999.tb09271.x

Nohara K, Zhang Y, Waraich RS, Laque A, Tiano JP, Tong J, Münzberg H, Mauvais-Jarvis F. 2011. Early-life exposure to testosterone programs the hypothalamic melanocortin system. *Endocrinology* **152**: 1661–1669. doi:10.1210/en.2010-1288

Nuñez JL, Huppenbauer CB, McAbee MD, Juraska JM, DonCarlos LL. 2003. Androgen receptor expression in the developing male and female rat visual and prefrontal cortex. *J Neurobiol* **56**: 293–302. doi:10.1002/neu.10236

O'Connell LA, Hofmann HA. 2011. The vertebrate mesolimbic reward system and social behavior network: a comparative synthesis. *J Comp Neurol* **519**: 3599–3639. doi:10.1002/cne.22735

Ogawa S, Lubahn DB, Korach KS, Pfaff DW. 1997. Behavioral effects of estrogen receptor gene disruption in male mice. *Proc Natl Acad Sci* **94**: 1476–1481. doi:10.1073/pnas.94.4.1476

Pérez SE, Chen E-Y, Mufson EJ. 2003. Distribution of estrogen receptor α and β immunoreactive profiles in the postnatal rat brain. *Brain Res Dev Brain Res* **145**: 117–139. doi:10.1016/S0165-3806(03)00223-2

Piekarski DJ, Johnson CM, Boivin JR, Thomas AW, Lin WC, Delevich K, Galarce EM, Wilbrecht L. 2017. Does puberty mark a transition in sensitive periods for plasticity in the associative neocortex? *Brain Res* **1654**: 123–144. doi:10.1016/j.brainres.2016.08.042

Priya R, Paredes MF, Karayannis T, Yusuf N, Liu X, Jaglin X, Graef I, Alvarez-Buylla A, Fishell G. 2018. Activity regulates cell death within cortical interneurons through a calcineurin-dependent mechanism. *Cell Rep* **22**: 1695–1709. doi:10.1016/j.celrep.2018.01.007

Raff MC, Barres BA, Burne JF, Coles HS, Ishizaki Y, Jacobson MD. 1993. Programmed cell death and the control of cell survival: lessons from the nervous system. *Science* **262**: 695–700. doi:10.1126/science.8235590

Raisman G, Field PM. 1971. Sexual dimorphism in the preoptic area of the rat. *Science* **173**: 731–733. doi:10.1126/science.173.3998.731

Schulz KM, Sisk CL. 2016. The organizing actions of adolescent gonadal steroid hormones on brain and behavioral development. *Neurosci Biobehav Rev* **70**: 148–158. doi:10.1016/j.neubiorev.2016.07.036

Scott N, Prigge M, Yizhar O, Kimchi T. 2015. A sexually dimorphic hypothalamic circuit controls maternal care and oxytocin secretion. *Nature* **525**: 519–522. doi:10.1038/nature15378

Shapiro LE, Leonard CM, Sessions CE, Dewsbury DA, Insel TR. 1991. Comparative neuroanatomy of the sexually dimorphic hypothalamus in monogamous and polygamous voles. *Brain Res* **541**: 232–240. doi:10.1016/0006-8993(91)91023-T

Shughrue PJ, Stumpf WE, MacLusky NJ, Zielinski JE, Hochberg RB. 1990. Developmental changes in estrogen receptors in mouse cerebral cortex between birth and postweaning: studied by autoradiography with 11β-methoxy-16α-[^{125}I]iodoestradiol. *Endocrinology* **126**: 1112–1124. doi:10.1210/endo-126-2-1112

Simerly RB, Swanson LW. 1988. Projections of the medial preoptic nucleus: a phaseolus vulgaris leucoagglutinin anterograde tract-tracing study in the rat. *J Comp Neurol* **270**: 209–242. doi:10.1002/cne.902700205

Cite this article as *Cold Spring Harb Perspect Biol* doi: 10.1101/cshperspect.a039099

Skene PJ, Henikoff S. 2017. An efficient targeted nuclease strategy for high-resolution mapping of DNA binding sites. *eLife* **6**: 1–35.

Södersten P. 1972. Mounting behavior in the female rat during the estrous cycle, after ovariectomy, and after estrogen or testosterone administration. *Horm Behav* **3**: 307–320. doi:10.1016/0018-506X(72)90020-7

Tasic B, Yao Z, Graybuck LT, Smith KA, Nguyen TN, Bertagnolli D, Goldy J, Garren E, Economo MN, Viswanathan S, et al. 2018. Shared and distinct transcriptomic cell types across neocortical areas. *Nature* **563**: 72–78. doi:10.1038/s41586-018-0654-5

Tsukahara S, Tsuda MC, Kurihara R, Kato Y, Kuroda Y, Nakata M, Xiao K, Nagata K, Toda K, Ogawa S. 2011. Effects of aromatase or estrogen receptor gene deletion on masculinization of the principal nucleus of the bed nucleus of the stria terminalis of mice. *Neuroendocrinology* **94**: 137–147. doi:10.1159/000327541

Vanderhaeghen P, Cheng HJ. 2010. Guidance molecules in axon pruning and cell death. *Cold Spring Harb Perspect Biol* **2**: a001859. doi:10.1101/cshperspect.a001859

van Veen JE, Kammel LG, Bunda PC, Shum M, Reid MS, Massa MG, Arneson D, Park JW, Zhang Z, Joseph AM, et al. 2020. Hypothalamic oestrogen receptor α establishes a sexually dimorphic regulatory node of energy expenditure. *Nat Metab* **2**: 351–363. doi:10.1038/s42255-020-0189-6

Waddington CH. 1940. *Organisers and genes*. Cambridge University Press, Cambridge.

Waters EM, Simerly RB. 2009. Estrogen induces caspase-dependent cell death during hypothalamic development. *J Neurosci* **29**: 9714–9718. doi:10.1523/JNEUROSCI.0135-09.2009

Wei D, Talwar V, Lin D. 2021. Neural circuits of social behaviors: innate yet flexible. *Neuron* **109**: 1600–1620. doi:10.1016/j.neuron.2021.02.012

Weisz J, Ward IL. 1980. Plasma testosterone and progesterone titers of pregnant rats, their male and female fetuses, and neonatal offspring. *Endocrinology* **106**: 306–316. doi:10.1210/endo-106-1-306

Welch JD, Kozareva V, Ferreira A, Vanderburg C, Martin C, Macosko EZ. 2019. Single-cell multi-omic integration compares and contrasts features of brain cell identity. *Cell* **177**: 1873–1887.e17. doi:10.1016/j.cell.2019.05.006

Wersinger SR, Sannen K, Villalba C, Lubahn DB, Rissman EF, De Vries GJ. 1997. Masculine sexual behavior is disrupted in male and female mice lacking a functional estrogen receptor α gene. *Horm Behav* **32**: 176–183. doi:10.1006/hbeh.1997.1419

Wu MV, Tollkuhn J. 2017. Estrogen receptor α is required in GABAergic, but not glutamatergic, neurons to masculinize behavior. *Horm Behav* **95**: 3–12. doi:10.1016/j.yhbeh.2017.07.001

Wu MV, Manoli DS, Fraser EJ, Coats JK, Tollkuhn J, Honda SI, Harada N, Shah NM. 2009. Estrogen masculinizes neural pathways and sex-specific behaviors. *Cell* **139**: 61–72. doi:10.1016/j.cell.2009.07.036

Yao Z, van Velthoven CTJ, Nguyen TN, Goldy J, Sedeno-Cortes AE, Baftizadeh F, Bertagnolli D, Casper T, Chiang M, Crichton K, et al. 2021. A taxonomy of transcriptomic cell types across the isocortex and hippocampal formation. *Cell* **184**: 3222–3241.e26. doi:10.1016/j.cell.2021.04.021

Yokosuka M, Okamura H, Hayashi S. 1995. Transient expression of estrogen receptor-immunoreactivity (ER-IR) in the layer V of the developing rat cerebral cortex. *Brain Res Dev Brain Res* **84**: 99–108. doi:10.1016/0165-3806(94)00161-R

Zhou JN, Hofman MA, Gooren LJ, Swaab DF. 1995. A sex difference in the human brain and its relation to transsexuality. *Nature* **378**: 68–70. doi:10.1038/378068a0

Zilkha N, Sofer Y, Kashash Y, Kimchi T. 2021. The social network: neural control of sex differences in reproductive behaviors, motivation, and response to social isolation. *Curr Opin Neurobiol* **68**: 137–151. doi:10.1016/j.conb.2021.03.005

Sex Differences in Circadian Rhythms

James C. Walton, Jacob R. Bumgarner, and Randy J. Nelson

Department of Neuroscience, Rockefeller Neuroscience Institute, West Virginia University, Morgantown, West Virginia 26506, USA

Correspondence: james.walton@hsc.wvu.edu

Sex as a biological variable is the focus of much literature and has been emphasized by the National Institutes of Health, in part, to remedy a long history of male-dominated studies in preclinical and clinical research. We propose that time-of-day is also a crucial biological variable in biomedical research. In common with sex differences, time-of-day should be considered in analyses and reported to improve reproducibility of studies and to provide the appropriate context to the conclusions. Endogenous circadian rhythms are present in virtually all living organisms, including bacteria, plants, invertebrates, and vertebrates. Virtually all physiological and behavioral processes display daily fluctuations in optimal performance that are driven by these endogenous circadian clocks; importantly, many of those circadian rhythms also show sex differences. In this review, we describe some of the documented sex differences in circadian rhythms.

Circadian rhythms are endogenous biological rhythms with periods of about 24 hours. Circadian rhythms persist in the absence of environmental cues; however, organisms use environmental cues, especially light, to entrain circadian rhythms precisely to the 24-hour solar day (Czeisler and Wright 1999). Synchronizing (or entraining) circadian rhythms to the solar day allows individuals to match physiological and behavioral responses with the appropriate temporal environmental conditions. Endogenous circadian rhythms are present in virtually all living organisms, including bacteria, plants, invertebrates, and vertebrates. Again, light is the most effective entraining agent, or *zeitgeber*. Among individuals of many vertebrate species, light stimulates intrinsically photosensitive retinal ganglion cells, which depolarize and synapse directly onto neurons in the suprachiasmatic nucleus (SCN) of the hypothalamus.

The master biological clock is located within the SCN where, dependent upon species, approximately 20,000–50,000 neurons maintain a transcriptional autoregulatory feedback loop. The molecular mechanism of the mammalian circadian clock has been reviewed in detail elsewhere (Partch et al. 2014). Virtually all cells have the clockwork mechanisms and are organized hierarchically throughout the body with the SCN serving as the master clock organizing all rhythms. The clockwork mechanism comprises an autoregulatory loop as the primary mechanism driving circadian rhythms; however, there is increasing evidence of additional processes, including posttranslational modifications (Gallego and Virshup 2007) and cAMP signaling (O'Neill et al.

2008), that are also critical to function. Time-of-day information, based on light intensity, is then relayed from the SCN to other brain regions, as well as to peripheral tissues, via neural and humoral pathways to provoke appropriate responses.

Sex as a biological variable is the focus of much literature and has been emphasized by the National Institutes of Health (NIH), in part, to remedy a long history of male-dominated studies in preclinical and clinical research (Beery and Zucker 2011; Zucker et al. 2021). Given the legions of sex differences in physiology and behavior, the exclusion of females from clinical and nonclinical research has likely had negative consequences for women's health (Beery and Zucker 2011; Zucker et al. 2021). Similarly, there are well-documented temporal differences in physiology and behavior that should be considered across all biological studies (Nelson et al. 2021).

We propose that time-of-day is also a crucial biological variable in biomedical research. In common with sex differences, time-of-day should be considered in analyses and reported to improve reproducibility of studies and to provide the appropriate context to the conclusions. Virtually all physiological and behavioral processes display daily fluctuations in optimal performance that are driven by endogenous circadian clocks; importantly, many of those circadian rhythms also show sex differences. Sex differences exist at multiple levels, from DNA to behavior, throughout the animal kingdom. In this article, we focus on sex differences in biological rhythms and how the neuroanatomical organization and hormonal milieu may transduce these differences or compensate for differences to normalize behavioral or physiological rhythms.

ANATOMICAL DIFFERENCES IN THE CIRCADIAN TIMING SYSTEM

There are three major afferent pathways through which zeitgebers can entrain the SCN; photic information via the retinohypothalamic tract (RHT) from the retina to the SCN, and nonphotic information transduced via the geniculohypothalamic tract (GHT) from the intergeniculate leaflet (IGL) to the SCN, or via direct projections from the dorsal and median raphe to the SCN. All of these afferent structures and the SCN express estrogen receptors (ERs) and androgen receptors (ARs) in various patterns (for reviews, see Bailey and Silver 2014; Yan and Silver 2016; Hatcher et al. 2020; Nicolaides and Chrousos 2020). Indeed, there are sex differences (and species-specific sex differences) in sex steroid receptor expression in the SCN (Iwahana et al. 2008), the retina (Wickham et al. 2000), the IGL (Horvath et al. 1999), and the raphe (Sheng et al. 2004); thus, gonadal and neurosteroids can directly affect the brain's master clock and its afferent pathways to influence the circadian system.

There are sex differences in how gonadal hormones affect both organization and modulation of SCN rhythmicity (Zucker et al. 1980; Albers 1981). Gonadal hormones can act directly on the SCN or indirectly via neurosteroid metabolites of gonadal steroids. Indeed, there is evidence that the SCN can synthesize neurosteroids such as progesterone, androsterone, and allo-tetrahydroxy-corticosterone (THDOC), which can alter SCN activity (Trachsel et al. 1996; Pinto and Golombek 1999). Although gonadal and neurosteroids and their receptors are in a position to directly modulate the SCN, sex differences in the effects of neurosteroids on the circadian system have yet to be fully investigated.

The SCN projects major efferents to over a dozen brain areas (Kriegsfeld and Silver 2006; Morin 2013), and these targets all express ERs and ARs in various combinations (Bailey and Silver 2014). Additionally, sex differences in function and anatomy of these SCN efferent target sites underlie sex differences in the hypothalamic pituitary adrenal (HPA) and hypothalamic pituitary gonadal (HPG) axes, as well as in sleep architecture and daily activity patterns (Semaan and Kauffman 2010; Morin 2013; Bailey and Silver 2014; Nicolaides and Chrousos 2020), which are discussed below.

PHYSIOLOGICAL DIFFERENCES IN THE CIRCADIAN TIMING SYSTEM

Hypothalamic Pituitary Gonadal (HPG) Axis

One of the most robust sex differences in circadian rhythmicity is found in circadian gating of the HPG axis. In female rodents, the SCN gates circadian timing of the preovulatory luteinizing hormone (LH) surge; however, estradiol concentration must be sufficiently high for the surge to occur (Christian and Moenter 2010; Williams and Kriegsfeld 2012). Similarly, there is a daily LH rhythm in women that also occurs at the onset of activity (Cahill et al. 1998). Males are unable to produce an LH surge, and it appears that this difference lies in sexually dimorphic population of kisspeptin neurons in the anteroventral periventricular nucleus (AVPV) (25 times more neurons in females than males), which project to GnRH neurons (for reviews, see Williams and Kriegsfeld 2012; Bailey and Silver 2014; Yan and Silver 2016). This neuroanatomical sex difference is a result of an organizational effect of gonadal steroids as developmental exposure to testosterone suppresses AVPV kisspeptin neuron numbers (Kauffman et al. 2007; Homma et al. 2009).

Hypothalamic Pituitary Adrenal (HPA) Axis

The SCN also regulates the HPA axis to affect glucocorticoid rhythms (Moore and Eichler 1972). At the level of the pituitary, this occurs presumably through direct and indirect actions of AVP neurons in the SCN and the PVN (Kalsbeek et al. 1992, 2010). The SCN can also regulate sensitivity of the adrenal cortex to ACTH in a circadian manner (Kaneko et al. 1981). There are also sex differences in HPA axis stress responsiveness (Handa et al. 1994, 2021), potentially due to sex differences in corticotropin-releasing factor (Bangasser and Wiersielis 2018) and/or liver X receptor α (Feillet et al. 2016), which may be downstream of sex-specific differences in AVP signaling in the SCN (Rohr et al. 2021). Some of these sex differences in stress responsiveness and the HPA axis can manifest in downstream physiological systems, such as the cardiovascular system.

Cardiovascular System

Various physiological features of the cardiovascular system are regulated by circadian rhythms, including heart rate, heart rate variability

(HRV), cardiovascular tone, angiogenesis, and vascular remodeling (Paschos and FitzGerald 2010). The circadian rhythms of these features are dictated by clock gene loops in the vascular endothelium, hormonal signals, and autonomic nervous signaling ultimately regulated by the SCN. As with many other circadian rhythms, several sex differences in cardiovascular circadian rhythms have been observed.

In general, females have higher resting and active heart rates than males, and this has been observed in humans to persist across the entire circadian day when examining the MESOR of heart rate (Hermida et al. 2002, 2007). For example, in a study in which ambulatory cardiovascular function was observed in referred patients, women had a slightly higher ambulatory heart rate across both the sleep and wake periods of the day (Ben-Dov et al. 2008). This observation persisted in another cohort study examining cardiovascular function in young and elderly populations (Stein et al. 1997). Another study demonstrated that the largest sex differences in heart rate occurred during the inactive phase in humans (Zhao et al. 2015).

Heart Rate Variability

Additional sex differences in cardiac function have been observed when examining time and frequency-domain indices of HRV across the day. Time-domain measures examine the intervals between specific components of polarization events during each cardiac cycle as measured by electrocardiograms. Frequency-domain measures assess the individual frequencies of functions of time-domain plots. For example, R-R intervals are a time-domain measurement, and the frequency-domain of the R-R interval can be determined through a Fourier transform of the R-R plotted over time (Shaffer and Ginsberg 2017).

In the time-domain, one study observed that men had greater R-R intervals across the entire circadian day (Bonnemeier et al. 2003). Men had increased standard deviations of NN intervals (SDNNs), standard deviation of the average NN intervals of each 5 min segment across a 24 h HRV recording (SDANN), and mean of

SDANN (SDNNi) across the entire day and greater root mean square of successive R-R interval differences (rMSSD) at night (Bonnemeier et al. 2003). Similar results were reported in a cohort study examining ~33-yr-old adults, where men had higher SDNN, SDANN, SDNNi, and average heart period in ms (AVGNN) than women (Stein et al. 1997). In contrast, a third study observed that women had lower R-R intervals during the inactive phase, but this difference was not statistically significant during the active phase (Extramiana et al. 1999). The same study reported that women had faster cardiac repolarization rates across the entire day, with the exception of the intervals between Q onset and T wave apex (Extramiana et al. 1999). In contrast, one study reported no difference in mean 24 h heart rates between male and female rhesus monkeys (Barger et al. 2010).

Sex differences in the frequency-domain of HRV across the day have also been observed. Women have lower LF/HF ratios than men during the circadian day (O'Connor et al. 2007). Another study reported similar results, where older men had greater LF/HF ratios across the entire day, but younger men only had a greater ratio during the inactive phase (Stein et al. 1997). Men display elevated LF during the active phase (Yamasaki et al. 1996) and across the entire day (Stein et al. 1997). Conversely, women displayed greater HF across the day (O'Connor et al. 2007), whereas another study reported that women have greater HF between 12:00–06:00 h (Yamasaki et al. 1996).

Last, depression differentially affects HRV across the day between women and men. In women, greater depressive scores were found to reduce the MESOR of circadian variation patterns of vagal activity, whereas the opposite effect was observed in men (Jarczok et al. 2018).

Blood Pressure Rhythms

In general, men tend to have higher systolic and diastolic blood pressure levels than women (Burt et al. 1995), an effect that persists across the day (Hermida et al. 2002, 2007, 2013). Ambulatory blood pressure monitoring in a cardiovascular clinic has also revealed that blood pressure levels

are lower in women than men across the day in an ambulatory setting (Ben-Dov et al. 2008). Other circadian-regulated aspects of blood pressure, such as dipper versus non-dipper patterns do not appear to be affected by biological sex (Ragot et al. 1999).

Rodent studies examining circadian differences in blood pressure have generated mixed results. No differences in arterial pressure were reported between male and female rats across the day (Sampson et al. 2008). Conversely, another study demonstrated that male C57Bl/6 and FVB/N mice display greater diastolic and systolic blood pressures across the day (Barsha et al. 2016). Among other potential mechanisms, differences in blood pressure may be driven by circadian rhythms in renal function, as constitutive renal *Bmal1* knockout in AQP2-Cre mice (C57Bl/6 background) led to reduced MESOR blood pressure in males but not females (Zhang et al. 2020).

Last, pharmacological treatment of blood pressure at different times of the day is affected by sex. Aspirin administration in the morning leads to elevated blood pressure in women, and aspirin administration in the evening leads to reduced blood pressure in both sexes, but a greater reduction is observed in women (Ayala and Hermida 2010).

Body Temperature Rhythms

Body temperature fluctuates across the day as a result of circadian regulated behavioral and physiological processes, such as variations in activity or metabolic function (Refinetti 2010). In general, men have lower body temperatures than women across the day. One study examining the effects of oral contraceptives on body temperature reported that men and naturally cycling women had lower MESOR body temperatures, greater temperature amplitudes, and lower nighttime-specific body temperature than women using oral contraceptives (Kattapong et al. 1995). Similar results were observed in another study, in which men had greater amplitudes in body temperature than women (Cain et al. 2010). Several studies have reported sex differences in phase angles of body temperature rhythms, with the nadir of body temperature during the inactive phase occurring 30 min (Baehr et al. 2000) to an hour earlier in women than in men (Cain et al. 2010). In free-running (i.e., not entrained) humans, women have been observed to have shorter intrinsic periods (taus) of body temperature rhythms (Wever 1984; Duffy et al. 2011). However, during internal desynchrony during free-running rhythms, the difference in body temperature taus between men and women is reported to disappear (Wever 1984).

Sex differences in body temperature have also been observed in other mammals. For example, female rhesus monkeys have lower body temperatures across the day with a greater phase angle than males, but with the acrophases being delayed rather than advanced as observed in humans (Barger et al. 2010). Potential mechanisms for sex differences in body temperature rhythms have been elucidated in mice. One study reported that differences in MESOR body temperatures between male and female mice are abolished after gonadectomy (GDX); GDX led to increased body temperatures in males and eliminated estrous-driven alterations of body temperature rhythms in females (Sanchez-Alavez et al. 2011).

Immune Function Rhythms

Sex differences in the immune system and function are discussed elsewhere in this collection (Moser 2021), and although there are well-studied circadian rhythms in the function of the immune system, few sex differences in the nature of these rhythms have been reported to date. For example, in humans, no differences in circadian rhythms were observed in the expression of Il-6 following lipopolysaccharide (LPS) stimulation; however, women mounted a more robust Il-6 response than men, which was associated with vagal tone and not with gonadal hormones (O'Connor et al. 2007). Female Lewis rats also display more robust responses to an immune challenge (ConA) than males; however, it is driven by biphasic increases in $CD8^+$ and MHC class II lymphocytes from the spleen at the end of both the light and dark phases (Griffin and Whitacre 1991). This sex-specific circadian difference in

immune response may underlie the sex differences in development of EAE (experimental allergic encephalomyelitis) in this strain of rats (Keith 1978), and reinforces the necessity of considering both sex and time-of-day as biological variables in future studies of autoimmune disorders.

SEX DIFFERENCES IN BEHAVIORAL RHYTHMS

Sleep–Wake Rhythms

Reports on sex differences in sleep–wake rhythms have been equivocal. Several studies reported that women have greater sleep fractions (Wever 1984) and greater sleep efficiency than men (Goel et al. 2005). These results are consistent with a survey where women stated greater ideal durations of sleep time in comparison to men (Tonetti et al. 2008). However, other self-reporting and survey studies have found no differences in sleep durations between men and women (Van Reen et al. 2013; Randler and Engelke 2019). Opposingly, one activity log study observed that adolescent women sleep less than men (Mathew et al. 2019). Differences between these studies may be a reflection of study design, age, and data-collection methods.

Differences in onset of sleep and activity have also been reported. Several studies reported that women have earlier wake times than men during adolescence (Mathew et al. 2019) and adulthood (Van Reen et al. 2013). One questionnaire study reported that women teachers have earlier bedtimes than men on weekdays, but not weekends (Randler and Engelke 2019), coinciding with a questionnaire study reporting later bedtimes for men than women (Tonetti et al. 2008). Conversely, several groups have reported no sex differences in time of activity onset (Baehr et al. 2000) or sleep onset (Cain et al. 2010). The latter study did report a sex difference in phase angle of sleep onset in relation to the onset of melatonin secretion (Cain et al. 2010). Last, young adult women (18–30 yr old) are reported to have earlier onsets of stage 1 and stage 2 sleep during the inactive phase than men (Goel et al. 2005).

Cite this article as *Cold Spring Harb Perspect Biol* doi: 10.1101/cshperspect.a039107

In terms of chronotype, men generally have later chronotypes than women; this difference begins around age 16 and disappears around age 50 (Roenneberg et al. 2004), consistent with other results (Adan and Natale 2002; Randler and Engelke 2019).

Similar sex differences in sleep and activity exist in rodents. Female rodents have greater variability in the onset of activity that corresponds with varying stages of the estrous cycle (Takahashi and Menaker 1980; Albers et al. 1981; Kuljis et al. 2013; Krizo and Mintz 2015). Female C57Bl/6J mice have longer αs (active phases) during constant darkness (Kuljis et al. 2013). Differences in α may reflect sex-differences in spontaneous firing rates in the dorsal SCN between Zeitgeber time (ZT) 4–6 (Kuljis et al. 2013). C57Bl/6 female mice have greater total percent time awake across the day, primarily during the active phase (Paul et al. 2006). It was also reported that females had reduced non–rapid eye movement (NREM) sleep and increased δ power during the active phase compared to males, and that these effects were driven by gonadal hormones (Paul et al. 2006). Indeed, circulating estrogen and aromatase activity at target sites in the circadian system during development and in adulthood have been implicated in sex differences in activity and circadian coupling in mice (for review, see Hatcher et al. 2020).

Although many of the effects of sex-specific estrogen described above are thought to drive the differences in activity and sleep described above, there is some indication that there may be differing effects of gonadal steroids and chromosomal sex on these parameters. One study addressed this possibility by using FCG (four core genotype) mice in which genetic sex is uncoupled from gonadal sex (Kuljis et al. 2013). Interestingly, the largest sex differences in this study were found after GDX in FCG mice. After GDX, activity levels rhythm power were reduced in both chromosomal sexes, but the XY mice had the greatest reduction, indicating that gonadal steroids had a greater effect on circadian rhythmicity than chromosomal sex (Kuljis et al. 2013). Because the FCG mice showed no sex differences in these measures prior to GDX, the authors concluded that the role of gonadal steroids was

to mask sex differences and normalize the behavior between sexes.

SEX DIFFERENCES IN THE EFFECTS OF DISRUPTION OF CIRCADIAN RHYTHMS

In humans, studies uncoupling endogenous circadian rhythms from the sleep–wake cycle using forced desynchrony and jet lag paradigms have revealed sex differences in cognition, affect, and physiology. In one study, compared to men, women had a higher amplitude of cognition performance and sleepiness after forced desynchrony (Santhi et al. 2016). Circadian misalignment after a 12 h phase shift increased circulating leptin in men, whereas women had decreased leptin coincident with increased ghrelin, resulting in altered food-type cravings between the sexes with no underlying difference in energy use (Qian et al. 2019). Although this study hints at a potential mechanism for sex differences in weight gain resulting from shift work, research in rodents has demonstrated that comparing circadian metabolic and transcriptional profiles between sexes is not straightforward. In mice, circadian transcriptomic profiling of the liver revealed that in females rhythms in expression were found in genes involved in cell signaling and protein transport, whereas in males the rhythmically expressed genes were involved in drug and steroid metabolism. There were also sex-specific effects of the microbiome on circadian transcription patterns (Weger et al. 2019). It may actually be a complex interaction among circadian rhythms in the gut microbiota, the circadian clock, and sex that feed back to regulate sex-specific differences in liver function and metabolism (Liang et al. 2015). Thus, it is apparent that future clinical metabolic studies must consider sex and time-of-day as critical biological variables.

Recent work in our laboratory has revealed striking sex differences in the effects of circadian disruption by exposure to dim light at night (dLAN) on many aspects of rodent physiology, behavior, and immune function. Although chronic mild circadian disruption by exposure to LAN (8 wk) in adulthood has similar effects on food intake resulting in obesity (Fonken et al.

2013; Aubrecht et al. 2015), LAN exposure during adolescence alters timing of food intake in male but not female mice, resulting in differential weight gain (Cissé et al. 2017). Brief disruption of circadian rhythms by as few as three nights of exposure to dLAN also alters brain physiology and behavior in a sex-specific manner in adult mice. Adult female mice displayed decreased anxiety-like behavior and had concurrent increases in *VEGFR1* and *IL-1B* expression in the brain, whereas males had reduced *BDNF* expression after three nights of dLAN exposure (Walker et al. 2020). These sex-specific effects of circadian disruption are not limited to adolescent and adult rodents however, because there appear to be transgenerational effects of circadian disruption that are sex-specific for both the sex of the parent and the sex of the offspring. In an experiment where adult male and female hamsters were exposed either to dark nights or to dLAN for 8 wk prior to conception, the male offspring of either sires or dams with preconception disrupted circadian rhythms had blunted immune responses and altered febrile response to an immune challenge compared to their female littermates (Cissé et al. 2020). However, only the female offspring of dams exposed to dLAN had enhanced bactericidal capacity of serum collected after an immune challenge, and none of these immunological effects were a result of altered maternal care (Cissé et al. 2020). This series of experiments makes it increasingly clear that even mild disruption of circadian rhythms has immediate and enduring sex-specific effects in animals at all stages of life, and these effects can be transgenerational in a sex-specific manner.

CONCLUDING REMARKS

Sex differences exist in circadian rhythms at all levels of analysis. In common with most areas of basic, clinical, and translational research, females have been understudied in circadian rhythm research. Furthermore, consideration of time-of-day as a biological variable is nearly nonexistent in most areas of research (Nelson et al. 2021). In this review, we have presented compelling evidence that there are critical sex differences in circadian rhythmicity in all aspects of biology across the life span. Furthermore, disruption of circadian rhythmicity by inappropriate exposure to LAN, shift work, or jet lag has sex-specific detrimental effects on physiology, behavior, and immune function, not only for those exposed, but potentially for future generations of their offspring as well. From a research standpoint, to improve reproducibility of studies and to provide the appropriate context to the conclusions, time-of-day must be reported and considered in experimental design and analyses of data. In common with eliminating male bias in research, unmasking time as a critical biological variable is long overdue.

ACKNOWLEDGMENTS

Preparation of this review was supported by National Institutes of Health Grants R01NS092388 and R21AT011238, as well as the National Institute of General Medical Sciences of the National Institutes of Health under Award No. 5U54GM104942-03. The content is solely the responsibility of the authors and does not necessarily represent the official views of the National Institutes of Health.

REFERENCES

*Reference is also in this collection.

Adan A, Natale V. 2002. Gender differences in morningness–eveningness preference. *Chronobiol Int* **19:** 709–720. doi:10.1081/CBI-120005390

Albers HE. 1981. Gonadal hormones organize and modulate the circadian system of the rat. *Am J Physiol* **241:** R62–R66.

Albers EE, Gerall AA, Axelson JF. 1981. Effect of reproductive state on circadian periodicity in the rat. *Physiol Behav* **26:** 21–25. doi:10.1016/0031-9384(81)90073-1

Aubrecht TG, Jenkins R, Nelson RJ. 2015. Dim light at night increases body mass of female mice. *Chronobiol Int* **32:** 557–560. doi:10.3109/07420528.2014.986682

Ayala D, Hermida R. 2010. Sex differences in the administration-time-dependent effects of low-dose aspirin on ambulatory blood pressure in hypertensive subjects. *Chronobiol Int* **27:** 345–362. doi:10.3109/07420521003624662

Baehr EK, Revelle W, Eastman CI. 2000. Individual differences in the phase and amplitude of the human circadian temperature rhythm: with an emphasis on morningness-eveningness. *J Sleep Res* **9:** 117–127. doi:10.1046/j.1365-2869.2000.00196.x

Bailey M, Silver R. 2014. Sex differences in circadian timing systems: implications for disease. *Front Neuroendocrinol* **35:** 111–139. doi:10.1016/j.yfrne.2013.11.003

Bangasser DA, Wiersielis KR. 2018. Sex differences in stress responses: a critical role for corticotropin-releasing factor. *Hormones (Athens)* **17:** 5–13. doi:10.1007/s42000-018-0002-z

Barger L, Hoban-Higgins T, Fuller C. 2010. Gender differences in the circadian rhythms of rhesus monkeys. *Physiol Behav* **101:** 595–600. doi:10.1016/j.physbeh.2010.06.002

Barsha G, Denton KM, Mirabito Colafella KM. 2016. Sex- and age-related differences in arterial pressure and albuminuria in mice. *Biol Sex Diff* **7:** 57. doi:10.1186/s13293-016-0110-x

Beery AK, Zucker I. 2011. Sex bias in neuroscience and biomedical research. *Neurosci Biobehav Rev* **35:** 565–572. doi:10.1016/j.neubiorev.2010.07.002

Ben-Dov I, Mekler J, Bursztyn M. 2008. Sex differences in ambulatory blood pressure monitoring. *Am J Med* **121:** 509–514. doi:10.1016/j.amjmed.2008.02.019

Bonnemeier H, Wiegand UK, Brandes A, Kluge N, Katus HA, Richardt G, Potratz J. 2003. Circadian profile of cardiac autonomic nervous modulation in healthy subjects: differing effects of aging and gender on heart rate variability. *J Cardiovasc Electrophysiol* **14:** 791–799. doi:10.1046/j.1540-8167.2003.03078.x

Burt V, Whelton P, Roccella E, Brown C, Cutler J, Higgins M, Horan M, Labarthe D. 1995. Prevalence of hypertension in the US adult population. Results from the Third National Health and Nutrition Examination Survey, 1988–1991. *Hypertension* **25:** 305–313. doi:10.1161/01.HYP.25.3.305

Cahill DJ, Wardle PG, Harlow CR, Hull MG. 1998. Onset of the preovulatory luteinizing hormone surge: diurnal timing and critical follicular prerequisites. *Fertil Steril* **70:** 56–59. doi:10.1016/S0015-0282(98)00113-7

Cain SW, Dennison CF, Zeitzer JM, Guzik AM, Khalsa SBS, Santhi N, Schoen MW, Czeisler CA, Duffy JF. 2010. Sex differences in phase angle of entrainment and melatonin amplitude in humans. *J Biol Rhythms* **25:** 288–296. doi:10.1177/0748730410374943

Christian CA, Moenter SM. 2010. The neurobiology of preovulatory and estradiol-induced gonadotropin-releasing hormone surges. *Endocr Rev* **31:** 544–577. doi:10.1210/er.2009-0023

Cissé YM, Peng J, Nelson RJ. 2017. Effects of dim light at night on food intake and body mass in developing mice. *Front Neurosci* **11:** 294. doi:10.3389/fnins.2017.00294

Cissé YM, Russart K, Nelson RJ. 2020. Exposure to dim light at night prior to conception attenuates offspring innate immune responses. *PLoS ONE* **15:** e0231140. doi:10.1371/journal.pone.0231140

Czeisler CA, Wright KP. 1999. *Influence of light on circadian rhythmicity in humans.* Marcel Dekker, New York.

Duffy JF, Cain SW, Chang AM, Phillips AJ, Münch MY, Gronfier C, Wyatt JK, Dijk DJ, Wright KP, Czeisler CA. 2011. Sex difference in the near-24-hour intrinsic period of the human circadian timing system. *Proc Natl Acad Sci* **108:** 15602–15608. doi:10.1073/pnas.1010666108

Extramiana F, Maison-Blanche P, Badilini F, Pinoteau J, Deseo T, Coumel P. 1999. Circadian modulation of QT rate dependence in healthy volunteers: gender and age differences. *J Electrocardiol* **32:** 33–43. doi:10.1016/S0022-0736(99)90019-5

Feillet C, Guérin S, Lonchampt M, Dacquet C, Gustafsson JA, Delaunay F, Teboul M. 2016. Sexual dimorphism in circadian physiology is altered in LXRα deficient mice. *PLoS ONE* **11:** e0150665. doi:10.1371/journal.pone.0150665

Fonken LK, Aubrecht TG, Meléndez-Fernández OH, Weil ZM, Nelson RJ. 2013. Dim light at night disrupts molecular circadian rhythms and increases body weight. *J Biol Rhythms* **28:** 262–271. doi:10.1177/0748730413493862

Gallego M, Virshup DM. 2007. Post-translational modifications regulate the ticking of the circadian clock. *Nat Rev Mol Cell Biol* **8:** 139–148. doi:10.1038/nrm2106

Goel N, Kim H, Lao R. 2005. Gender differences in polysomnographic sleep in young healthy sleepers. *Chronobiol Int* **22:** 905–915. doi:10.1080/07420520500263235

Griffin AC, Whitacre CC. 1991. Sex and strain differences in the circadian rhythm fluctuation of endocrine and immune function in the rat: implications for rodent models of autoimmune disease. *J Neuroimmunol* **35:** 53–64. doi:10.1016/0165-5728(91)90161-Y

Handa RJ, Burgess LH, Kerr JE, O'Keefe JA. 1994. Gonadal steroid hormone receptors and sex differences in the hypothalamo-pituitary-adrenal axis. *Horm Behav* **28:** 464–476. doi:10.1006/hbeh.1994.1044

* Handa RJ, Sheng JA, Castellanos EA, Templeton NH, McGivern RF. 2021. Sex differences in acute neuroendocrine responses to stressors in rodents and humans. *Cold Spring Harb Perspect Med* doi:10.1101/cshperspect.a039081

Hatcher KM, Royston SE, Mahoney MM. 2020. Modulation of circadian rhythms through estrogen receptor signaling. *Eur J Neurosci* **51:** 217–228. doi:10.1111/ejn.14184

Hermida R, Ayala D, Fernández J, Mojón A, Alonso I, Calvo C. 2002. Modeling the circadian variability of ambulatorily monitored blood pressure by multiple-component analysis. *Chronobiol Int* **19:** 461–481. doi:10.1081/CBI-120002913

Hermida R, Ayala D, Portaluppi F. 2007. Circadian variation of blood pressure: the basis for the chronotherapy of hypertension. *Adv Drug Deliv Rev* **59:** 904–922. doi:10.1016/j.addr.2006.08.003

Hermida R, Ayala D, Mojón A, Fontao M, Chayán L, Fernández J. 2013. Differences between men and women in ambulatory blood pressure thresholds for diagnosis of hypertension based on cardiovascular outcomes. *Chronobiol Int* **30:** 221–232. doi:10.3109/07420528.2012.701487

Homma T, Sakakibara M, Yamada S, Kinoshita M, Iwata K, Tomikawa J, Kanazawa T, Matsui H, Takatsu Y, Ohtaki T, et al. 2009. Significance of neonatal testicular sex steroids to defeminize anteroventral periventricular kisspeptin neurons and the GnRH/LH surge system in male rats. *Biol Reprod* **81:** 1216–1225. doi:10.1095/biolreprod.109.078311

Horvath TL, Diano S, Sakamoto H, Shughrue PJ, Merchenthaler I. 1999. Estrogen receptor β and progesterone receptor mRNA in the intergeniculate leaflet of the female rat. *Brain Res* **844:** 196–200. doi:10.1016/S0006-8993(99)01759-X

Iwahana E, Karatsoreos I, Shibata S, Silver R. 2008. Gonadectomy reveals sex differences in circadian rhythms and suprachiasmatic nucleus androgen receptors in mice. *Horm Behav* **53**: 422–430. doi:10.1016/j.yhbeh.2007.11.014

Jarczok MN, Aguilar-Raab C, Koenig J, Kaess M, Borniger JC, Nelson RJ, Hall M, Ditzen B, Thayer JF, Fischer JE. 2018. The heart's rhythm "n" blues: sex differences in circadian variation patterns of vagal activity vary by depressive symptoms in predominantly healthy employees. *Chronobiol Int* **35**: 896–909. doi:10.1080/07420528.2018.1439499

Kalsbeek A, Buijs RM, van Heerikhuize JJ, Arts M, van der Woude TP. 1992. Vasopressin-containing neurons of the suprachiasmatic nuclei inhibit corticosterone release. *Brain Res* **580**: 62–67. doi:10.1016/0006-8993(92)909 27-2

Kalsbeek A, Fliers E, Hofman MA, Swaab DF, Buijs RM. 2010. Vasopressin and the output of the hypothalamic biological clock. *J Neuroendocrinol* **22**: 362–372. doi:10.1111/j.1365-2826.2010.01956.x

Kaneko M, Kaneko K, Shinsako J, Dallman MF. 1981. Adrenal sensitivity to adrenocorticotropin varies diurnally. *Endocrinology* **109**: 70–75. doi:10.1210/endo-109-1-70

Kattapong KR, Fogg LF, Eastman CI. 1995. Effect of sex, menstrual cycle phase, and oral contraceptive use on circadian temperature rhythms. *Chronobiol Int* **12**: 257–266. doi:10.3109/07420529509057274

Kauffman AS, Gottsch ML, Roa J, Byquist AC, Crown A, Clifton DK, Hoffman GE, Steiner RA, Tena-Sempere M. 2007. Sexual differentiation of *Kiss1* gene expression in the brain of the rat. *Endocrinology* **148**: 1774–1783. doi:10.1210/en.2006-1540

Keith AB. 1978. Sex difference in Lewis rats in the incidence of recurrent experimental allergic encephalomyelitis. *Nature* **272**: 824–825. doi:10.1038/272824a0

Kriegsfeld LJ, Silver R. 2006. The regulation of neuroendocrine function: timing is everything. *Horm Behav* **49**: 557–574. doi:10.1016/j.yhbeh.2005.12.011

Krizo JA, Mintz EM. 2015. Sex differences in behavioral circadian rhythms in laboratory rodents. *Front Endocrinol* **5**: 234. doi:10.3389/fendo.2014.00234

Kuljis DA, Loh DH, Truong D, Vosko AM, Ong ML, McClusky R, Arnold AP, Colwell CS. 2013. Gonadal- and sex-chromosome-dependent sex differences in the circadian system. *Endocrinology* **154**: 1501–1512. doi:10.1210/en.2012-1921

Liang X, Bushman FD, FitzGerald GA. 2015. Rhythmicity of the intestinal microbiota is regulated by gender and the host circadian clock. *Proc Natl Acad Sci* **112**: 10479–10484. doi:10.1073/pnas.1501305112

Mathew G, Hale L, Chang A. 2019. Sex moderates relationships among school night sleep duration, social jetlag, and depressive symptoms in adolescents. *J Biol Rhythms* **34**: 205–217. doi:10.1177/0748730419828102

Moore RY, Eichler VB. 1972. Loss of a circadian adrenal corticosterone rhythm following suprachiasmatic lesions in the rat. *Brain Res* **42**: 201–206. doi:10.1016/0006-8993 (72)90054-6

Morin LP. 2013. Neuroanatomy of the extended circadian rhythm system. *Exp Neurol* **243**: 4–20. doi:10.1016/j.expneurol.2012.06.026

∗ Moser A. 2021. Sex differences in autoimmune/inflammatory disease. *Cold Spring Harb Perspect Med* doi:10.1101/cshperspect.a039172

Nelson RJ, Bumgarner JB, Walker WH, DeVries AC. 2021. Time-of-day as a critical biological variable. *Neurosci Biobehav Rev* **127**: 740–746. doi:10.1016/j.neubiorev.2021.05.017

Nicolaides NC, Chrousos GP. 2020. Sex differences in circadian endocrine rhythms: clinical implications. *Eur J Neurosci* **52**: 2575–2585. doi:10.1111/ejn.14692

O'Connor M, Motivala S, Valladares E, Olmstead R, Irwin M. 2007. Sex differences in monocyte expression of IL-6: role of autonomic mechanisms. *Am J Physiol Regul Integr Comp Physiol* **293**: R145–R151. doi:10.1152/ajpregu.00752.2006

O'Neill JS, Maywood ES, Chesham JE, Takahashi JS, Hastings MH. 2008. cAMP-dependent signaling as a core component of the mammalian circadian pacemaker. *Science* **320**: 949–953. doi:10.1126/science.1152506

Partch CL, Green CB, Takahashi JS. 2014. Molecular architecture of the mammalian circadian clock. *Trends Cell Biol* **24**: 90–99. doi:10.1016/j.tcb.2013.07.002

Paschos G, FitzGerald G. 2010. Circadian clocks and vascular function. *Circ Res* **106**: 833–841. doi:10.1161/CIRCRESAHA.109.211706

Paul KN, Dugovic C, Turek FW, Laposky AD. 2006. Diurnal sex differences in the sleep-wake cycle of mice are dependent on gonadal function. *Sleep* **29**: 1211–1223. doi:10.1093/sleep/29.9.1211

Pinto FT, Golombek DA. 1999. Neuroactive steroids alter the circadian system of the Syrian hamster in a phase-dependent manner. *Life Sci* **65**: 2497–2504. doi:10.1016/S0024-3205(99)00516-0

Qian J, Morris CJ, Caputo R, Wang W, Garaulet M, Scheer FAJL. 2019. Sex differences in the circadian misalignment effects on energy regulation. *Proc Natl Acad Sci* **116**: 23806–23812. doi:10.1073/pnas.1914003116

Ragot S, Herpin D, Siché J, Ingrand P, Mallion J. 1999. Autonomic nervous system activity in dipper and non-dipper essential hypertensive patients. What about sex differences? *J Hypertens* **17**: 1805–1811. doi:10.1097/00004872-199917121-00004

Randler C, Engelke J. 2019. Gender differences in chronotype diminish with age: a meta-analysis based on morningness/chronotype questionnaires. *Chronobiol Int* **36**: 888–905. doi:10.1080/07420528.2019.1585867

Refinetti R. 2010. The circadian rhythm of body temperature. *Front Biosci* **15**: 564–594. doi:10.2741/3634

Roenneberg T, Kuehnle T, Pramstaller PP, Ricken J, Havel M, Guth A, Merrow M. 2004. A marker for the end of adolescence. *Curr Biol* **14**: R1038–R1039. doi:10.1016/j.cub.2004.11.039

Rohr KE, Telega A, Savaglio A, Evans JA. 2021. Vasopressin regulates daily rhythms and circadian clock circuits in a manner influenced by sex. *Horm Behav* **127**: 104888. doi:10.1016/j.yhbeh.2020.104888

Sampson A, Widdop R, Denton K. 2008. Sex-differences in circadian blood pressure variations in response to chronic angiotensin II infusion in rats. *Clin Exp Pharmacol Physiol* **35**: 391–395. doi:10.1111/j.1440-1681.2008.04884.x

Sanchez-Alavez M, Alboni S, Conti B. 2011. Sex- and age-specific differences in core body temperature of C57Bl/6 mice. *Age (Omaha)* **33:** 89–99. doi:10.1007/s11357-010-9164-6

Santhi N, Lazar AS, McCabe PJ, Lo JC, Groeger JA, Dijk D-J. 2016. Sex differences in the circadian regulation of sleep and waking cognition in humans. *Proc Natl Acad Sci* **113:** E2730–E2739. doi:10.1073/pnas.1521637113

Semaan SJ, Kauffman AS. 2010. Sexual differentiation and development of forebrain reproductive circuits. *Curr Opin Neurobiol* **20:** 424–431. doi:10.1016/j.conb.2010.04.004

Shaffer F, Ginsberg J. 2017. An overview of heart rate variability metrics and norms. *Front Public Health* **5:** 258. doi:10.3389/fpubh.2017.00258

Sheng Z, Kawano J, Yanai A, Fujinaga R, Tanaka M, Watanabe Y, Shinoda K. 2004. Expression of estrogen receptors (α, β) and androgen receptor in serotonin neurons of the rat and mouse dorsal raphe nuclei; sex and species differences. *Neurosci Res* **49:** 185–196. doi:10.1016/j.neures.2004.02.011

Stein PK, Kleiger RE, Rottman JN. 1997. Differing effects of age on heart rate variability in men and women. *Am J Cardiol* **80:** 302–305. doi:10.1016/S0002-9149(97)00350-0

Takahashi JS, Menaker M. 1980. Interaction of estradiol and progesterone: effects on circadian locomotor rhythm of female golden hamsters. *Am J Physiol* **239:** R497–R504.

Tonetti L, Fabbri M, Natale V. 2008. Sex difference in sleeptime preference and sleep need: a cross-sectional survey among Italian pre-adolescents, adolescents, and adults. *Chronobiol Int* **25:** 745–759. doi:10.1080/07420520802394191

Trachsel L, Dodt HU, Zieglgänsberger W. 1996. The intrinsic optical signal evoked by chiasm stimulation in the rat suprachiasmatic nuclei exhibits GABAergic day–night variation. *Eur J Neurosci* **8:** 319–328. doi:10.1111/j.1460-9568.1996.tb01216.x

Van Reen E, Sharkey KM, Roane BM, Barker D, Seifer R, Raffray T, Bond TL, Carskadon MA. 2013. Sex of college students moderates associations among bedtime, time in bed, and circadian phase angle. *J Biol Rhythms* **28:** 425–431. doi:10.1177/0748730413511771

Walker WH II, Borniger JC, Gaudier-Diaz MM, Hecmarie Meléndez-Fernández O, Pascoe JL, Courtney DeVries A, Nelson RJ. 2020. Acute exposure to low-level light at night is sufficient to induce neurological changes and depressive-like behavior. *Mol Psychiatry* **25:** 1080–1093. doi:10.1038/s41380-019-0430-4

Weger BD, Gobet C, Yeung J, Martin E, Jimenez S, Betrisey B, Foata F, Berger B, Balvay A, Foussier A, et al. 2019. The mouse microbiome is required for sex-specific diurnal rhythms of gene expression and metabolism. *Cell Metab* **29:** 362–382.e8. doi:10.1016/j.cmet.2018.09.023

Wever R. 1984. Sex differences in human circadian rhythms: intrinsic periods and sleep fractions. *Experientia* **40:** 1226–1234. doi:10.1007/BF01946652

Wickham LA, Gao J, Toda I, Rocha EM, Ono M, Sullivan DA. 2000. Identification of androgen, estrogen and progesterone receptor mRNAs in the eye. *Acta Ophthalmol Scand* **78:** 146–153. doi:10.1034/j.1600-0420.2000.078002146.x

Williams WP 3rd, Kriegsfeld LJ. 2012. Circadian control of neuroendocrine circuits regulating female reproductive function. *Front Endocrinol (Lausanne)* **3:** 60.

Yamasaki Y, Kodama M, Matsuhisa M, Kishimoto M, Ozaki H, Tani A, Ueda N, Ishida Y, Kamada T. 1996. Diurnal heart rate variability in healthy subjects: effects of aging and sex difference. *Am J Physiol* **271:** H303–H310.

Yan L, Silver R. 2016. Neuroendocrine underpinnings of sex differences in circadian timing systems. *J Steroid Biochem Mol Biol* **160:** 118–126. doi:10.1016/j.jsbmb.2015.10.007

Zhang D, Jin C, Obi IE, Rhoads MK, Soliman RH, Sedaka RS, Allan JM, Tao B, Speed JS, Pollock JS, et al. 2020. Loss of circadian gene *Bmal1* in the collecting duct lowers blood pressure in male, but not female, mice. *Am J Physiol Renal Physiol* **318:** F710–F719. doi:10.1152/ajprenal.00364.2019

Zhao R, Li D, Zuo P, Bai R, Zhou Q, Fan J, Li C, Wang L, Yang X. 2015. Influences of age, gender, and circadian rhythm on deceleration capacity in subjects without evident heart diseases. *Ann Noninvasive Electrocardiol* **20:** 158–166. doi:10.1111/anec.12189

Zucker I, Fitzgerald KM, Morin LP. 1980. Sex differentiation of the circadian system in the golden hamster. *Am J Physiol* **238:** R97–R101.

* Zucker I, Prendergast BJ, Beery AK. 2021. Pervasive neglect of sex differences in biomedical research. *Cold Spring Harb Perspect Biol* doi:10.1101/cshperspect.a039156.

Sex Differences in Acute Neuroendocrine Responses to Stressors in Rodents and Humans

Robert J. Handa,[1,3] Julietta A. Sheng,[1] Emily A. Castellanos,[1] Hayley N. Templeton,[1] and Robert F. McGivern[2]

[1]Department of Biomedical Sciences, Colorado State University, Fort Collins, Colorado 80523, USA

[2]Department of Psychology, San Diego State University, San Diego, California 92120, USA

Correspondence: rmcgivern@sdsu.edu

Sex differences in the neuroendocrine response to acute stress occur in both animals and humans. In rodents, stressors such as restraint and novelty induce a greater activation of the hypothalamic-pituitary-adrenal axis (HPA) in females compared to males. The nature of this difference arises from steroid actions during development (organizational effects) and adulthood (activational effects). Androgens decrease HPA stress responsivity to acute stress, while estradiol increases it. Androgenic down-regulation of HPA responsiveness is mediated by the binding of testosterone (T) and dihydrotestosterone (DHT) to the androgen receptor, as well as the binding of the DHT metabolite, 3β-diol, to the β form of the estrogen receptor (ERβ). Estradiol binding to the α form of the estrogen receptor (ERα) increases HPA responsivity. Studies of human sex differences are relatively few and generally employ a psychosocial paradigm to measure stress-related HPA activation. Men consistently show greater HPA reactivity than women when being evaluated for achievement. Some studies have found greater reactivity in women when being evaluated for social performance. The pattern is inconsistent with rodent studies but may involve the differential nature of the stressors employed. Psychosocial stress is nonphysical and invokes a significant degree of top-down processing that is not easily comparable to the types of stressors employed in rodents. Gender identity may also be a factor based on recent work showing that it influences the neural processing of positive and negative emotional stimuli independent of genetic sex. Comparing different types of stressors and how they interact with gender identity and genetic sex will provide a better understanding of sex steroid influences on stress-related HPA reactivity.

Hans Selye (1950) introduced the term "stress" to describe a unidimensional reflex arc initiated by physiological and psychological factors that trigger endocrine and autonomic responses to threatening stimuli. This formed the basis of his General Adaptation Syndrome that emphasized the role of elevated levels of adrenal glucocorticoids (GCs) as an essential link between stress and medical illness. The reflexive nature of the response was later modified

---

[3]Deceased.

Cite this article as Cold Spring Harb Perspect Biol doi: 10.1101/cshperspect.a039081

by studies in humans and primates showing that elevated cortisol levels were strongly related to the degree of situational control and anxiety, as opposed to overall physiological reactivity (Mason 1975). This perspective led to the view that neuroendocrine responses to stress reflect the impact of psychosocial factors versus arousal. Further modifications to the reflexive concept arose from studies demonstrating dissociations between elevated cortisol levels and self-report of anxiety or stress levels in humans due to individual differences, as well as differences between men and women (Kudielka and Kirschbaum 2005; Reschke-Hernández et al. 2017).

The current model emphasizes individual differences in psychological stress reactivity operating within a conceptual framework consisting of an integrated set of distinct physiological, emotional, and cognitive networks (Buigs and Van Eden 2000; Goldstein et al. 2010). The physiological networks include neuroendocrine and autonomic responses that increase glucose availability, regulate cardiovascular tone, and suppress gonadal hormone release and immune responses. Emotional networks include the limbic system and amygdala, which coordinate motivational state with reflexive behavioral responses related to defensive or aggressive behaviors. Cognitive networks include the prefrontal and cingulate cortices, which serve to focus selective attention on social context and planning strategies for restoring behavioral equilibrium. The integration of these networks resides in hypothalamic circuitry that coordinates the systemic response to stressors according to physiological needs and behavioral goals. The networks are similar in rodents and humans, but the evolutionary expansion of the human telencephalon provides the cognitive network a more dominant and flexible role in psychological stress reactivity.

The fundamental work on physiological and behavioral reactivity to stress comes from animal studies conducted over the past century. These have set the stage for translational studies of psychological stress reactivity in humans and its role in disease and mental disorders (Frankenhaeuser 1996; Cohen et al. 2007; Zorn et al. 2017). However, studies that compared physio-

logical and behavioral stress reactivity in men and women were uncommon for most of the twentieth century (Kajantie and Phillips 2006). This omission largely stemmed from a general assumption that sex differences in human physiological responses to stress were small and their relationship to illness was unlikely to extend beyond reproductive systems.

Over the past 20 years, that assumption has been challenged by a growing literature showing a variety of nonreproductive medical problems that can be exacerbated by stress interacting with the genetic sex of the individual. Among these are susceptibility to cardiovascular and autoimmune diseases, as well as affective disorders such as anxiety and depression (Maeng and Milad 2015; Rubinow and Schmidt 2019). However, there is still a significant gap in our mechanistic understanding of how gonadal steroids interact with genetic sex in human stress (Shors 2016; Helpman et al. 2017).

This article first considers mechanisms regulating the activation and feedback inhibition of the hypothalamic-pituitary-adrenal axis (HPA) in rodents, with a focus on how steroids modulate this system to establish sex differences in adulthood and during development. The discussion is followed by the consideration of functional sex differences in human stress activity and general approaches currently used to study stress-related HPA activation.

REGULATION OF THE HPA: ANIMAL STUDIES IN RODENTS

Selye's pioneering work showing that the adrenal glands of female rats are larger than in males provided the first evidence suggesting a potential sex difference in the physiological stress response (Selye 1937). This sex difference was later correlated with a higher rate of basal and stress-induced adrenal GC secretion in females, an effect that is mediated by estrogen (Kitay 1961, 1964). It is now well established in rodents that females exhibit greater diurnal fluctuations in plasma GCs and greater GC secretion in response to physical or psychological stressors (Bielohuby et al. 2007; Goel et al. 2014; Zuloaga et al. 2020). The magnitude of this sex difference in rodents is remarkable, with basal and stress-reactive levels of corticosterone (CORT) up to 1.5 to 2 times greater in females than males (Aloisi et al. 1998).

In mammals, adrenal GC secretion is controlled by the hypothalamus. The HPA represents a cascade of neural and humoral signals driven by actual or perceived changes in the environment (Handa and Weiser 2014). In response to changes that may affect physiological homeostasis, the HPA is activated through groups of neuropeptide-expressing neurons located within the paraventricular nucleus (PVN) of the hypothalamus. Corticotropin-releasing hormone (CRH) expressing neurons, located in the parvocellular subdivisions of the PVN, are critical for this response. The release of CRH into the hypophyseal portal system stimulates the release of adrenocorticotropic hormone (ACTH) from anterior pituitary corticotrophs, which subsequently stimulates the adrenal cortex to produce and release GCs into the general circulation. Circulating GCs then feed back at the anterior pituitary, hypothalamus, and higher brain areas to negatively regulate further secretion and, importantly, to alter stress-related behaviors (Herman et al. 1998, 2012).

Functional Sex Differences in the HPA

Sex differences in HPA function have been consistently reported in the literature, and some of the mechanisms underlying these differences have now been identified. In rats, the ACTH and CORT response of females (Viau et al. 2005; Iwasaki-Sekino et al. 2009; Heck and Handa 2019a) is characterized by a greater and prolonged secretion of ACTH and GCs indicating both enhanced stimulus reactivity, as well as reduced negative feedback (Handa et al. 1994; Babb et al. 2013). Changes in negative feedback have been tracked to changes in gonadal steroid levels (Heck et al. 2020).

Importantly, the HPA response to stressors should not be considered detrimental as GC hormones act in a largely beneficial fashion in the short term. Acute rises in GCs augment physiological functions involved in the fight or flight reaction, enhance cognition, and limit functions unnecessary for an immediate stress response (e.g., reproduction, immune function, digestion) (Lupien et al. 2002; Charmandari et al. 2005; Yuen et al. 2009). In rats, gonadectomy (GDX) of males and females minimizes the sex difference in CORT secretion showing that the response is partially due to sex differences in circulating gonadal hormones (Heck and Handa 2019a). Moreover, hormone replacement to GDX animals reinstates the size of the sex difference found in intact animals (Heck and Handa 2019a). Thus, numerous studies have demonstrated that estradiol (E2) enhances, whereas testosterone (T) treatment inhibits HPA reactivity (Viau and Meaney 1996; Heck and Handa 2019b). However, some studies have found that E2 can reduce HPA reactivity, indicating the complexity of this response may be tied to other factors besides just hormone presence. Whether these contrasting results are due to hormone dose, type or length of treatment exposure, or duration of GDX has not yet been explored in detail. As a result, the mechanisms by which these hormones act to influence HPA function are not completely resolved (Viau et al. 1999; Handa et al. 2013; Oyola et al. 2016).

In rats, the most important hypothalamic releasing factors for pituitary ACTH secretion are CRH and arginine vasopressin (AVP). Both factors are potent ACTH secretagogues with AVP potentiating ACTH-releasing activity of CRH several-fold both in vivo and in vitro (Rivier and Vale 1983; Young et al. 2007). Moreover, females have significantly greater numbers of CRH-immunoreactive (CRH-ir) neuronal cell bodies within the PVN than do males (Stinnett et al. 2015), and also exhibit greater diurnal variation of CRH immunoreactivity in these neurons than males (Critchlow et al. 1963; Smith and Norman 1987; Handa and McGivern 2017). These morphological data are consistent with physiological data showing a greater stress-responsive activation of the HPA axis in females (Seale et al. 2004).

Studies have shown greater female expression of AVP and CRH mRNA in the PVN, and more ACTH precursor (POMC) in anterior pituitary following restraint compared to males (Viau et al. 2005; Babb et al. 2013). However, sex differences in AVP expression have been reported for other brain areas, such as the bed nucleus of the stria terminalis (BNST), lateral septum, medial preoptic area (MPOA), and amygdala (Rood and De Vries 2011; DiBenedictis et al. 2017), where the number of AVP neurons is much greater in males than females (de Vries et al. 2008). Given that AVP augments the actions of CRH at the anterior pituitary and can be coreleased with CRH at the median eminence, greater extra-PVN levels in males likely indicate that sex differences in AVP are related to behavioral effects of central AVP neurotransmission rather than HPA regulation.

Sex Differences in Negative Feedback Regulation of the HPA

The stimulatory limb of the HPA axis is kept under continuous check by feedback inhibition (Sapolsky et al. 2000). Negative feedback consists of GC sensitive inputs from a number of upstream regions including the hippocampus, BNST, prefrontal cortex, and others (Herman et al. 2012). One factor underlying sex differences in GC secretion involves the intensity of feedback inhibition. The negative feedback action of GCs is mediated by two corticosteroid receptor types: type I or mineralocorticoid receptor (MR) and type II or GC receptor (GR). Both receptor types reside within cells of the hypothalamus, hippocampus, and anterior pituitary gland, as well as other brain areas (Reul and de Kloet 1985). Both receptors appear to be involved in differing aspects of feedback regulation.

MRs possess a very high affinity for corticosteroids and, as a result, are predominantly occupied at basal levels of hormone secretion (Reul and de Kloet 1985). GRs, in contrast, exhibit an approximately 10-fold lower affinity for GCs and, therefore, are occupied following GC elevations (Reul and de Kloet 1985), such as after a stress-induced rise in GCs. Although MR has a higher affinity for corticosteroids than GR, these receptors function together to return stress-responsive elevations in corticosteroids to baseline (Goel et al. 2014). Assuming receptor number corresponds to hormone sensitivity, sex differences in the HPA may arise as a result of sex differences

in receptor number, indicating greater negative feedback sensitivity and lower basal and stress-activated hormone secretion. Conversely, lower receptor levels correspond to weaker negative feedback and, thus, higher basal and stress-responsive hormone secretion (Kolber and Muglia 2009; Solomon et al. 2015). The concentration of GR and MR and the function of GR and MR are lower in females than males in the hypothalamus, hippocampus, and other brain regions (Solomon et al. 2015). This pattern is consistent with that of other tissues such as the thymus and liver (Herman et al. 2016). Given similar levels of hormone, GR- or MR-mediated functions (such as autoregulation of the receptor) are less sensitive to GC modulation in females than males (Endres et al. 1979), indicating that other factors modulate GC function.

E2 Modulation of the HPA

The stage of the estrous cycle affects the female neuroendocrine response to stress as reflected in the plasma levels of corticosterone. Basal and stress-responsive corticosteroid levels are highest on proestrus when E2 levels are correspondingly high (Viau and Meaney 1991; Herman et al. 2016; Heck and Handa 2019a). This may be a result of impaired negative feedback regulation (Heck and Handa 2019a). Hormone replacement studies show that E2 treatment of GDX female and male rats enhances basal and stress-responsive secretion of ACTH and corticosteroids (Seale et al. 2004; Lund et al. 2006; Figueiredo et al. 2007; Weiser and Handa 2009). The actual mechanism(s) whereby E2 enhances stress-responsive ACTH and CORT secretion is not completely resolved. At the hypothalamic level, increases in HPA activation could be due to increased stimulation of PVN neurons, or reductions in inhibitory tone (Heck and Handa 2019b). Such possibilities are suggested by data showing increased expression of stress-inducible c-fos mRNA in the PVN following E2-treatment (Larkin et al. 2010). The prolonged stress-responsive secretion of ACTH and corticosteroids, coupled with elevated baseline secretion of ACTH and corticosteroids in females following E2-treatment, also suggests reduced HPA negative feedback mechanisms. Experimental manipulation of CORT secretion by E2 shows a similar pattern, where CORT secretion is suppressed in GDX females administered by the synthetic GC, dexamethasone (de Souza et al. 2019). This suppressive effect of E2 may be mediated by reduced γ-aminobutyric acid (GABA)-ergic inhibition as GAD67-positive neurons express estrogen receptor α (ERα) in the peri-PVN region (Weiser and Handa 2009).

Androgen Modulation of HPA Function

In comparison to E2, a large body of literature suggests androgens exert mainly inhibitory actions on the HPA axis (Viau and Meaney 1996; Sheng et al. 2021b). GDX of adult male rats increases CORT and ACTH responses to stressors and, correspondingly, the expression of c-fos in PVN neurons (Handa et al. 2013; Rosinger et al. 2019). Hormone replacement in GDX male rats with T or the nonaromatizable androgen, dihydrotestosterone (DHT), returns stress-responsive CORT and ACTH back to the levels found in intact males (Lund et al. 2006; Williamson and Viau 2008). Treatment of GDX animals with DHT also inhibits the stress-induced rise in PVN c-fos mRNA demonstrating that T effects are independent of T aromatization to E2 (Williamson and Viau 2008). Importantly, inhibition of 5α reductase in males causes a rise in basal and postrestraint corticosteroid secretion (Handa et al. 2013; Heck and Handa 2019a).

Although androgens can inhibit HPA function and reduce CRH-ir in the PVN (Heck and Handa 2019a), androgen receptors (ARs) are not found in CRH or AVP neurons within the rodent PVN (Bingaman et al. 1994; Heck and Handa 2019b). ARs have been reported in the PVN, but these AR-ir neurons are located in subdivisions of the PVN that project to spinal cord and brainstem autonomic nuclei (Bingham et al. 2011; Heck and Handa 2019b). Consequently, the assumption is that AR regulation of the HPA occurs trans-synaptically. Local application of T to the BNST and MPOA inhibits HPA responses to stress in adult male rats (Williamson and Viau 2008). However, local application of DHT to the PVN of males shows that

DHT can directly affect PVN neurons through a mechanism not involving AR, but rather through estrogen receptor β (ERβ) (Lund et al. 2006; Handa et al. 2009). Local PVN application of DHT effectively inhibited HPA function. Moreover, local application of 5α-androstane-3β, 17β-diol (3β-diol), a metabolite of DHT that binds ERβ, was as effective as DHT in inhibiting HPA function as was PVN application of ERβ-selective agonists (Lund et al. 2006). Consistent with this finding, the effects of both DHT and 3β-diol can be blocked by ER antagonists, but not an AR antagonist treatment (Lund et al. 2006). Because 3β-diol can bind ERβ with moderate affinity, these data support the hypothesis that DHT can be metabolized to 3β-diol, a ligand that preferentially binds ERβ, and that ERβ acts to inhibit HPA function. These effects of DHT metabolites on ERβ also occur in females (Kudwa et al. 2014), although circulating DHT levels are much lower in females than males (Fig. 1; Handelsman et al. 2018).

Organizational Actions of Gonadal Steroids Underlying Sex-Biases in HPA Function

Exposure to varying levels of gonadal steroids during perinatal life programs the brain to set up sex biases in the HPA. Such "organizational" actions permanently modify the morphology and neural circuitry of the brain (Sheng et al. 2021a). In male rodents, there are two important surges of T during pre- and postnatal development that are crucial for defeminization and masculinization of the rat brain: during late gestation days 18–19 and 2–4 h after parturition (Corbier et al. 1978; McGivern et al. 1988). Male adult rats that were GDX at birth showed elevated stress-induced ACTH and CORT secretions, increased PVN c-Fos activation, and reduced AR expression in BNST and medial amygdala nucleus, similar to females. Neonatal T-replacement reversed these effects (Bingham and Viau 2008). Moreover, T treatment of neonatal female rats leads to decreased HPA reactivity in adulthood, suggesting organizational effects by gonadal steroids that persist into adulthood (Seale et al. 2005). Chen et al. (2014) used a novel knockout mouse model to demonstrate

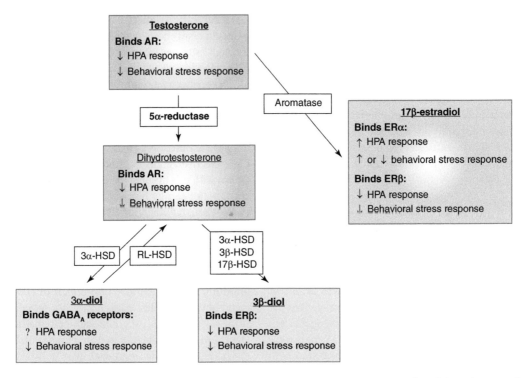

Figure 1. Effects of testosterone (T) and its metabolites on hypothalamic-pituitary-adrenal (HPA) axis and behavioral stress responses. This figure describes enzymes involved in the conversion of T and its metabolites and predicted effects produced by binding androgen receptors (ARs), estrogen receptor α (ERα), estrogen receptor β (ERβ), and γ-aminobutyric acid (GABA) receptors. Binding of ARs or ERβ is expected to decrease the HPA axis and behavioral stress responses. In contrast, actions at ERα increase the HPA axis response to stress. Effects of 3α-diol on the HPA axis are currently unknown. (HSD) Hydroxysteroid dehydrogenase, (3α-diol) 5α-androstane-3α, 17β-diol, (3β-diol) 5α-androstane-3β, 17β-diol, (RL-HSD) 11-*cis*-retinol dehydrogenase-like 3α-HSD. (The figure and legend are reprinted from Zuloaga et al. 2020 under the terms of a Creative Commons Attribution 4.0 International License.)

that male mice with a testicular feminization mutation (Tfm), due to a dysfunctional *AR*, exhibit a more female-like HPA response to stressors. Such data indicate that T's modulation of the HPA in adulthood is AR-dependent as demonstrated by earlier studies using other Tfm models (Zuloaga et al. 2008, 2011). However, T can also be converted to E2 by the aromatase enzyme, which appears to be important for the organizational actions of T on the HPA response to stress. Indeed, because neonatal castration affects adult behavioral and CORT responses to stress in Tfm rats (Zuloaga et al. 2011), which have a dysfunctional AR, these organizational effects of T cannot be mediated by AR. In support of a role for aromatization, Bingham et al. (2012) showed elevated stress-induced CORT secretions and in-

creased PVN c-Fos expression in neonate males that were implanted with a slow-release capsule containing an aromatase inhibitor 12 h after birth. Such a response is more typical of females and, therefore, these data implicate an alternate, ER-mediated organizational effect of T on the masculinization of the HPA.

SEX DIFFERENCES IN STRESS-INDUCED ACTIVATION OF THE HPA: HUMAN STUDIES

The pattern of stress-induced HPA activation in men and women discussed below exhibits some notable differences from sex differences observed in animals (Donner and Lowry 2013). Before addressing these differences, it is important to note differences in the perceptual nature of psychological stress between rodents and humans. The induction of psychological stress in animals relies strongly on physical experience (Altemus 2006). This includes procedures that have a physical or reflexive behavioral basis, such as restraint, swimming, shock, or species-specific behaviors like the inherent avoidance of open spaces in rodents (Bangasser and Wicks 2017; Rincón-Cortés et al. 2019). In contrast, human studies of neuroendocrine stress reactivity generally employ psychosocial procedures that have no physical component. Cross-species comparisons of what constitutes "stress" is also complicated by the evolutionary expansion of the human prefrontal cortex, including its reciprocal connections with the anterior cingulate cortex and amygdala, which allows more complex socioemotional analysis of internal and external stimuli (McEwen et al. 2015). This appraisal system is also influenced by hormone level changes associated with the menstrual cycle (Goel et al. 2014). Few animal studies have examined stress reactivity across the estrous cycle. Thus, human stress studies of nonclinical populations are generally designed around complex socioemotional-type stressors that are not easily modeled in animals.

The first programmatic study of human sex differences in stress-induced HPA activation was initiated by Frankenhaeuser (1996) and colleagues more than 50 years ago. In both laboratory and real-world situations, they found that men exhibited greater cortisol and autonomic responses than women when the stress involved external evaluation related to social standing and/or academic achievement. Kirschbaum et al. (1992, 1999) later formalized this psychosocial approach by developing a standardized experimental procedure known as the Trier Social Stress Test (TSST). Today, the TSST is the most widely used method for studying sex differences in human stress reactivity (Allen et al. 2017).

In the TSST procedure, participants are given a 3-min period to prepare a short interview-type speech to be presented before an unresponsive audience of evaluators. This is followed by 5 min for presentation of the speech, at the end of which participants are surprised with a 5-min mental arithmetic test. Subjective stress level is assessed with self-reported measures of stress, task difficulty, and anxiety level. Neuroendocrine activation is generally measured by salivary cortisol levels and many studies also include heart rate as an indicator of autonomic reactivity. Whereas the TSST is widely viewed as a general measure of psychosocial stress, it is important to note that its approach is weighted toward inducing achievement-related stress because it includes cognitive skills that are evaluated in a social context.

TSST studies consistently demonstrate greater male HPA activation during the stress period compared to females (Kirschbaum et al. 1999; Kudielka and Kirschbaum 2005; Uhart et al. 2006; Stephens et al. 2016; Reschke-Hernández et al. 2017). Interestingly, males also exhibited greater HPA reactivity than females in anticipation of experiencing the task. This finding suggests the sex difference in stress-induced cortisol levels arises from input to the PVN from higher levels related to perceptual evaluation of the experience. A limited amount of data indicate that the sex-related pattern is not related to pituitary sensitivity to CRH as no sex differences in cortisol levels were found following CRH infusion under nonstress conditions or during physical exercise stress (Kirschbaum et al. 1999).

Although males consistently show higher cortisol responses in the TSST, the size of the sex difference can vary depending upon the phase of the menstrual cycle. Numerous studies have reported female cortisol responses are smaller during the follicular phase compared to the luteal phase, while a few have found no difference (Kajantie and Phillips 2006; Stevens and Hamann 2012). Oral contraceptives blunt the female cortisol response or have no effect (Kirschbaum et al. 1999; Bouma et al. 2009; Cornelisse et al. 2011; Klumbies et al. 2014; Barel et al. 2018). Overall, the impact of hormonal differences between cycle phases is variable across TSST studies, somewhat contradictory, and associated with relatively small sample sizes (Stevens and Hamann 2012). A clearer picture emerges from a study of 798 men and women by Herbison et al. (2016). During the stress period, male cortisol levels were higher than female levels during all phases of the menstrual cycle. The largest sex difference occurred in the follicular phase, with smaller or similar differences observed at ovulation and the luteal phase. This pattern suggests that elevated E and P levels enhance the perception of stress in women.

The different cortisol response pattern indicates that men and women do not perceive the psychosocial stress of the TSST in the same way. This inference is consistent with established sex differences in the overall behavioral response to psychosocial threat. On average, men are more likely to exhibit direct physical or verbal aggression when socially challenged, compared with the tendency of women to employ indirect measures related to social exclusion (Benenson et al. 2011). Thus, the achievement aspect of the TSST procedure may contribute more to the male experience of stress threat and HPA activation than to the female experience.

Stroud et al. (2002) provided some support for this interpretation in a well-controlled study that used two conditions to induce psychosocial stress. The first was an achievement situation conceptually analogous to the TSST. Participants were assessed for verbal and arithmetic skills after being told the purpose of the study was to examine the relationship between physiological responses and intelligence. To enhance psychosocial stress, two confederates were employed in the social assessment phase who knew most of the answers. This ensured that participants performed poorly. The second condition involved social rejection and used the same two confederates, where the participant and the confederates were told that the purpose of the study was to better understand how individuals get to know one another. They would discuss two different topics while the experimenter videotaped the interactions. The confederates were trained to use subtle social and verbal skills to socially exclude the participant over the course of the discussion. Female participants in both conditions were balanced for menstrual cycle phases. The results showed a marked interaction between sex and condition in the HPA response. In the achievement condition, there was a twofold rise in male cortisol levels compared to baseline but no change in female levels. In the social

rejection condition, there was a twofold rise in female levels but no change in male levels. Interestingly, men and women rated both conditions equally stressful.

Wang et al. (2007) incorporated functional magnetic resonance imaging (fMRI) into a variation of the TSST to examine neural activation patterns. Low and high stress levels were induced by having participants perform simple and complicated mental arithmetic tasks while being evaluated by researchers. Performance of males and females was similar. However, males perceived the condition as more stressful than females, whereas females rated the same condition as significantly more difficult than males. In this study, no stress-induced sex differences were observed in heart rate or cortisol responses, which may reflect the modifications made to accommodate the fMRI. Nevertheless, robust sex differences were observed in activation of the cortical stress response network. Male activation of the right prefrontal cortex was increased compared to females, while activation of the left orbitofrontal cortex was suppressed. Females showed greater activation in limbic areas that included the hippocampus, insula, and anterior cingulate cortex. Sex differences in cortical activation patterns were large enough that a discriminant statistical analysis of male and female patterns correctly classified the sex of the participant with 94% accuracy.

Although HPA responsivity in the TSST is smaller in women, females show enhanced limbic activation in some areas compared to men when processing negative emotional stimuli. Stimuli associated with negative emotions induced greater amygdala activity in women in contrast to greater amygdala responses to positive emotions in men (Kogler et al. 2015). Some studies have not found this pattern, but lack of controls for menstrual cycle phase may be a factor (Schienle et al. 2005; Stevens and Hamann 2012; Seo et al. 2017). Overall, the differential pattern in cortical and limbic activity may reflect greater depth of emotional processing in women, which may also contribute to their higher level of social cognition (Kret and de Gelder 2012).

Goldstein et al. (2010) demonstrated E2's effect on stress-related neural activation using a within-subject design that used fMRI to scan neural activity of women at two phases of the menstrual cycle. Stress was induced by having participants view visual stimuli with established negative valence. Women were scanned in the early follicular phase when E2 levels are low, and again in the midcycle phase around the time of ovulation when E2 levels are high, but before the large rise of P during the mid- to late luteal phase. Men were included in the study for comparison. Results showed few sex differences in cortical activation when women were in the early follicular phase. However, during the midcycle phase, the neural response of women was significantly less than men in the limbic areas that included the amygdala, hippocampus, anterior cingulate gyrus, medial prefrontal cortex, and orbital prefrontal cortex.

Males and females differ in the resting state functional connectivity of the basolateral amygdala, the region that stimulates fear responses in rats and humans, suggesting that the difference in the response to emotional stimuli is inherent. Females show greater basolateral connections with lateral frontal and striatal regions, while males show greater connectivity with medial frontal regions (Engman et al. 2016; McGlade et al. 2020). This pattern indicates that females have a hormonal capacity to regulate the stress response differently from males, resulting in a more efficient and effective connectivity that translates stress responses to subjective awareness (Ordaz and Luna 2012).

Because of the historical exclusion of sex in studies of stress reactivity, especially in nonclinical populations (Zucker et al. 2021), the current literature is limited, and leaves open important questions related to the broader issue of individual differences, some of which involve the definition of sex in research. It is generally understood that "sex" refers to sex chromosomes, while the term "gender" is defined by sociocultural expectations of behaviors and can denote a range of identities that do not have to correspond to established ideas of male and female based on sex chromosomes. Yet, the terms sex and gender are often used interchangeably in scientific papers and in studies of sex differences. The human studies cited in this chapter classified partici-

pants on the basis of genetic sex, leaving open the question of whether gender identification is a superseding influence on stress reactivity that is distinct from genetic sex.

Yuan et al. (2021) recently addressed part of this question in a study of men and women that included the constructs of gender and sex as independent variables. They examined selective attention using event-related potentials (ERPs) associated with visual neural processing of neutral stimuli as well as positive and negative stimuli of varying levels of emotional intensity. They employed an "oddball paradigm" with an infrequent standardized neutral stimulus shown randomly. Participants responded on each trial as to whether the stimulus matched the standardized neutral stimulus. Each participant was assessed for their gender identity, with their "masculine" and "feminine" profile scores included as an independent factor in addition to biological sex. Gender profiles were derived from a Chinese Sex Role Inventory modeled on the original Bem Sex Role Inventory (Bem 1981; Keyes 1984; Qian et al. 2000). The ERP response to both positive and negative stimuli revealed a significantly larger P100 component in both hemispheres for individuals with feminine gender roles compared to masculine. This effect was not observed when individuals were compared by genetic sex. Participants classified as feminine, regardless of biological sex, exhibited greater sensitivity to the valence of both positive and negative stimuli compared to participants classified as masculine. This feminine classification was associated with significantly longer reaction times to discriminate the standardized neutral stimulus compared to those classified as masculine. Biological males and females showed no sex difference in reaction time. A better understanding of steroid regulation of human HPA reactivity will emerge from future studies that incorporate gender identification as a defining variable in addition to genetic sex.

CONCLUSIONS

Studies of sex differences in the acute neuroendocrine response of rodents to stress show a role for both organizational and activational effects of gonadal steroid actions that regulate the HPA. Organizational effects of androgens induce a long-term decrease in HPA stress reactivity. Activational effects of androgens generally down-regulate HPA stress sensitivity in the rodent, whereas estrogens increase it. Androgenic down-regulation involves T and DHT binding to the AR, as well as the DHT metabolite, 3β-diol, binding to ERβ. Estrogen stimulation of HPA sensitivity is mediated through E2 binding to ERα. Sex differences are also found in hypothalamic expression of CRH and AVP, as well the concentration of GRs in brain areas that regulate HPA feedback.

Much less is known about the mechanisms of sex steroid regulation of HPA activity in humans. Men and women also exhibit differences in HPA reactivity following acute psychological stress, but the pattern differs from rodents. The species difference may reflect a lack of experimental equivalency between the stressors used in rodents versus the psychosocial stressors employed in men and women.

Understanding differences between men and women in their neuroendocrine stress responses is obviously a complex process and likely involves individual differences operating at multiple levels. Currently, our limited knowledge does not provide a clear path toward explaining how these differences relate to sex differences observed in stress-related clinical disorders. The task is growing more complex with the recognition of a role for gender and personality as additional factors associated with differences in structural and functional connectivity in brain regions that include stress networks (Burke et al. 2017; Nostro et al. 2017; Xin et al. 2017). Future studies integrating these factors into studies of hormonal and genetic influences on stress responsiveness in preclinical and clinical populations will be important for developing translational approaches to understanding and treating stress-related mental disorders and diseases.

ACKNOWLEDGMENTS

Bob Handa passed away in August 2021 after an extended illness. He was a dear friend and collaborator for more than 40 years. A great mind, a big heart, and a wonderful teacher who loved fishing almost as much as research. He is greatly missed. Bob had a strong commitment to mentoring a younger generation and specifically asked that this review be completed to honor the intellectual contributions made to it by so many of his students over the past three decades. —Robert F. McGivern.

REFERENCES

*Reference is also in this collection.

Allen AP, Kennedy PJ, Dockray S, Cryan JF, Dinan TG, Clarke G. 2017. The Trier Social Stress Test: principles and practice. *Neurobiol Stress* **6:** 113–126. doi:10.1016/j.ynstr.2016.11.001

Aloisi AM, Ceccarelli I, Lupo C. 1998. Behavioural and hormonal effects of restraint stress and formalin test in male and female rats. *Brain Res Bull* **47:** 57–62. doi:10.1016/S0361-9230(98)00063-X

Altemus M. 2006. Sex differences in depression and anxiety disorders: potential biological determinants. *Horm Behav* **50:** 534–538. doi:10.1016/j.yhbeh.2006.06.031

Babb JA, Masini CV, Day HEW, Campeau S. 2013. Sex differences in activated corticotropin-releasing factor neurons within stress-related neurocircuitry and hypothalamic-pituitary-adrenocortical axis hormone following restraint in rats. *Neurosci* **234:** 40–52. doi:10.1016/j.neuroscience.2012.12.051

Bangasser DA, Wicks B. 2017. Sex-specific mechanisms for responding to stress. *J Neurosci Res* **95:** 75–82. doi:10.1002/jnr.23812

Barel E, Abu-Shkara R, Colodner R, Masalha R, Mahagna L, Zemel OC, Cohen A. 2018. Gonadal hormones modulate the HPA-axis and the SNS in response to psychosocial stress. *J Neurosci Res* **96:** 1388–1397. doi:10.1002/jnr.24259

Bem SL. 1981. Gender schema theory: a cognitive account of sex typing. *Psychol Rev* **88:** 354–364. doi:10.1037/0033-295X.88.4.354

Benenson JF, Markovits H, Thompson ME, Wrangham RW. 2011. Under threat of social exclusion, females exclude more than males. *Psychol Sci* **22:** 538–544. doi:10.1177/0956797611402511

Bielohuby M, Herbach N, Wanke R, Maser-Gluth C, Beuschlein F, Wolf E, Hoeflich A. 2007. Growth analysis of the mouse adrenal gland from weaning to adulthood: time- and gender-dependent alterations of cell size and number in the cortical compartment. *J Clin Endocrinol Metab* **293:** E139–E146.

Bingaman EW, Magnuson DJ, Gray TS, Handa RJ. 1994. Androgen inhibits the increases in hypothalamic corticotropin-releasing hormone (CRH) and CRH-immunoreactivity following gonadectomy. *Neuroendocrinology* **59:** 228–234. doi:10.1159/000126663

Bingham B, Viau V. 2008. Neonatal gonadectomy and adult testosterone replacement suggest an involvement of limbic arginine vasopressin and androgen receptors in the organization of the hypothalamic-pituitary-adrenal axis.

Cite this article as *Cold Spring Harb Perspect Biol* doi: 10.1101/cshperspect.a039081

Endocrinology **149:** 3581–3591. doi:10.1210/en.2007-1796

Bingham B, Myung C, Innala L, Gray M, Anonuevo A, Viau V. 2011. Androgen receptors in the posterior bed nucleus of the stria terminalis increase neuropeptide expression and the stress-induced activation of the paraventricular nucleus of the hypothalamus. *Neuropsychopharmacology* **36:** 1433–1443. doi:10.1038/npp.2011.27

Bingham B, Wang NXR, Innala L, Viau V. 2012. Postnatal aromatase blockade increases c-fos mRNA responses to acute restraint stress in adult male rats. *Endocrinology* **153:** 1603–1608. doi:10.1210/en.2011-1749

Bouma EM, Riese H, Ormel J, Verhulst FC, Oldehinkel AJ. 2009. Adolescents' cortisol responses to awakening and social stress; effects of gender, menstrual phase and oral contraceptives. The TRAILS study. *Psychoneuroendocrinology* **34:** 884–893. doi:10.1016/j.psyneuen.2009.01.003

Buijs RM, Van Eden CG. 2000. The integration of stress by the hypothalamus, amygdala, and prefrontal cortex: balance between the autonomic nervous system and the neuroendocrine systems. *Prog Br Res* **126:** 117–131. doi:10.1016/S0079-6123(00)26011-1

Burke SM, Manzouri AH, Savic I. 2017. Structural connections in the brain in relation to gender identity and sexual orientation. *Sci Rep* **7:** 17954. doi:10.1038/s41598-017-17352-8

Charmandari E, Tsigos C, Chrousos G. 2005. Endocrinology of the stress response. *Annu Rev Physiol* **67:** 259–284. doi:10.1146/annurev.physiol.67.040403.120816

Chen CV, Brummet JL, Lonstein JS, Jordan CL, Breedlove SM. 2014. New knockout model confirms a role for androgen receptors in regulating anxiety-like behaviors and HPA response in mice. *Horm Behav* **65:** 211–218. doi:10.1016/j.yhbeh.2014.01.001

Cohen S, Janicki-Deverts D, Miller GE. 2007. Psychological stress and disease. *JAMA* **298:** 1685–1687. doi:10.1001/jama.298.14.1685

Corbier P, Kerdelhue B, Picon R, Roffi J. 1978. Changes in testicular weight and serum gonadotropin and testosterone levels before, during, and after birth in the perinatal rats. *Endocrinology* **103:** 1985–1991. doi:10.1210/endo-103-6-1985

Cornelisse S, van Stegeren AH, Joëls M. 2011. Implications of psychosocial stress on memory formation in a typical male versus female student sample. *Psychoneuroendocrinology* **36:** 569–578. doi:10.1016/j.psyneuen.2010.09.002

Critchlow V, Liebelt RA, Bar-Sela M, Mountcastle W, Lipscomb HS. 1963. Sex difference in resting pituitary-adrenal function in the rat. *Am J Physiol* **205:** 807–815. doi:10.1152/ajplegacy.1963.205.5.807

de Souza CF, Stopa LRS, Santos GF, Takasumi LCN, Martins AB, Garnica-Siqueira MC, Ferreira RN, de Andrade FG, Leite CM, Zaia DAM, et al. 2019. Estradiol protects against ovariectomy-induced susceptibility to the anabolic effects of glucocorticoids in rats. *Life Sci* **218:** 185–196. doi:10.1016/j.lfs.2018.12.037

de Vries GJ, Jardon M, Reza M, Rosen GJ, Immerman E, Forger NG. 2008. Sexual differentiation of vasopressin innervation of the brain: cell death versus phenotypic differentiation. *Endocrinology* **149:** 4632–4637. doi:10.1210/en.2008-0448

DiBenedictis BT, Nussbaum ER, Cheung HK, Veenema AH. 2017. Quantitative mapping reveals age and sex differences in vasopressin, but not oxytocin, immunoreactivity in the rat social behavior neural network. *J Comp Neurol* **525:** 2549–2570. doi:10.1002/cne.24216

Donner NC, Lowry CA. 2013. Sex differences in anxiety and emotional behavior. *Pflugers Arch* **465:** 601–626. doi:10.1007/s00424-013-1271-7

Endres DB, Milholland RJ, Rosen F. 1979. Sex differences in the concentrations of glucocorticoid receptors in rat liver and thymus. *J Endocrinol* **80:** 21–26. doi:10.1677/joe.0.0800021

Engman J, Linnman C, Van Dijk KR, Milad MR. 2016. Amygdala subnuclei resting-state functional connectivity sex and estrogen differences. *Psychoneuroendocrinology* **63:** 34–42. doi:10.1016/j.psyneuen.2015.09.012

Figueiredo HF, Ulrich-Lai YM, Choi DC, Herman JP. 2007. Estrogen potentiates adrenocortical responses to stress in female rats. *Am J Physiol Endocrinol Metab* **292:** E1173–E1182. doi:10.1152/ajpendo.00102.2006

Frankenhaeuser M. 1996. Stress and gender. *European Review* **4:** 313–327. doi:10.1002/(SICI)1234-981X(199610)4:4<313::AID-EURO139>3.0.CO;2-Z

Goel N, Workman JL, Lee TT, Innala L, Viau V. 2014. Sex differences in the HPA axis. *Compr Physiol* **4:** 1121–1155. doi:10.1002/cphy.c130054

Goldstein JM, Jerram M, Abbs B, Whitfield-Gabrieli S, Makris N. 2010. Sex differences in stress response circuitry activation dependent on female hormonal cycle. *J Neurosci* **30:** 431–438. doi:10.1523/JNEUROSCI.3021-09.2010

Handa RJ, McGivern RF. 2017. Stress response: sex differences. In *Reference module in neuroscience and biobehavioral psychology*, pp. 1–6. Elsevier, Philadelphia, PA. http://dx.doi.org/10.1016/B978-0-12-809324-5.02865-0

Handa RJ, Weiser MJ. 2014. Gonadal steroid hormones and the hypothalamo-pituitary-adrenal axis. *Front Neuroendocrinol* **35:** 197–220. doi:10.1016/j.yfrne.2013.11.001

Handa RJ, Burgess LH, Kerr JE, O'Keefe JA. 1994. Gonadal steroid hormone receptors and sex differences in the hypothalamo-pituitary-adrenal axis. *Horm Behav* **28:** 464–476. doi:10.1006/hbeh.1994.1044

Handa RJ, Weiser MJ, Zuloaga DG. 2009. A role for the androgen metabolite, 5α-androstane-3β, 17β-diol, in modulating oestrogen receptor β-mediated regulation of hormonal stress reactivity. *J Neuroendocrinol* **44:** 451–458. doi:10.1111/j.1365-2826.2009.01840.x

Handa RJ, Kudwa AE, Donner NC, McGivern RF, Brown R. 2013. Central 5-α reduction of testosterone is required for testosterone's inhibition of the hypothalamo-pituitary-adrenal axis response to restraint stress in adult male rats. *Brain Res* **1529:** 74–82. doi:10.1016/j.brainres.2013.07.021

Handelsman DJ, Hirschberg AL, Bermon S. 2018. Circulating testosterone as the hormonal basis of sex differences in athletic performance. *Endocr Rev* **39:** 803–829. doi:10.1210/er.2018-00020

Heck AL, Handa RJ. 2019a. Sex differences in the hypothalamic-pituitary-adrenal axis' response to stress: an important role for gonadal hormones. *Neuropsychopharmacology* **44:** 45–58. doi:10.1038/s41386-018-0167-9

Heck AL, Handa RJ. 2019b. Androgens drive sex biases in hypothalamic corticotropin-releasing hormone gene expression after adrenalectomy of mice. *Endocrinology* **160:** 1757–1770. doi:10.1210/en.2019-00238

Heck AL, Sheng JA, Miller MA, Stover SA, Bales NJ, Tan SM, Daniels RM, Fleury TK, Handa RJ. 2020. Social isolation alters hypothalamic pituitary adrenal axis activity after chronic variable stress in male C57BL/6 mice. *Stress* **23:** 457–465. doi:10.1080/10253890.2020.1733962

Helpman L, Zhu X, Suarez-Jimenez B, Lazarov A, Monk C, Neria Y. 2017. Sex differences in trauma-related psychopathology: a critical review of neuroimaging literature (2014–2017). *Curr Psychiatry Rep* **19:** 104. doi:10.1007/s11920-017-0854-y

Herbison CE, Henley D, Marsh J, Atkinson H, Newnham JP, Matthews SG, Lye SJ, Pennell CE. 2016. Characterization and novel analyses of acute stress response patterns in a population-based cohort of young adults: influence of gender, smoking, and BMI. *Stress* **19:** 139–150. doi:10.3109/10253890.2016.1146672

Herman JP, Dolgas CM, Carlson SL. 1998. Ventral subiculum regulates hypothalamo-pituitary-adrenocorticoid and behavioural responses to cognitive stressors. *Neuroscience* **86:** 449–459. doi:10.1016/S0306-4522(98)00055-4

Herman JP, McKlveen JM, Solomon MB, Carvalho-Netto E, Myers B. 2012. Neural regulation of the stress response: glucocorticoid feedback mechanisms. *Braz J Med Biol Res* **45:** 292–298. doi:10.1590/S0100-879X2012007500041

Herman JP, McKlveen JM, Ghosal S, Kopp B, Wulsin A, Makinson R, Scheimann J, Myers B. 2016. Regulation of the hypothalamic-pituitary-adrenocortical stress response. *Compr Physiol* **6:** 603–621. doi:10.1002/cphy.c150015

Iwasaki-Sekino A, Mano-Otagiri A, Ohata H, Yamauchi N, Shibasaki T. 2009. Gender differences in corticotropin and corticosterone secretion and corticotropin-releasing factor mRNA expression in the paraventricular nucleus of the hypothalamus and the central nucleus of the amygdala in response to footshock stress or psychological stress in rats. *Psychoneuroendocrinology* **34:** 226–237. doi:10.1016/j.psyneuen.2008.09.003

Kajantie E, Phillips DI. 2006. The effects of sex and hormonal status on the physiological response to acute psychosocial stress. *Psychoneuroendocrinology* **31:** 151–178. doi:10.1016/j.psyneuen.2005.07.002

Keyes S. 1984. Measuring sex-role stereotypes: attitudes among Hong Kong Chinese adolescents and the development of the Chinese sex-role inventory. *Sex Roles* **10:** 129–140. doi:10.1007/BF00287752

Kirschbaum C, Wüst S, Faig HG, Hellhammer DH. 1992. Heritability of cortisol responses to human corticotropin-releasing hormone, ergometry, and psychological stress in humans. *J Clin Endocrinol Metab* **75:** 1526–1530. doi:10.1210/jcem.75.6.1464659

Kirschbaum C, Kudielka BM, Gaab J, Schommer NC, Hellhammer DH. 1999. Impact of gender, menstrual cycle phase, and oral contraceptives on the activity of the hypothalamus-pituitary-adrenal axis. *Psychosom Med* **61:** 154–162. doi:10.1097/00006842-199903000-00006

Kitay JI. 1961. Sex differences in adrenal cortical secretion in the rat. *Endocrinology* **68:** 818–824. doi:10.1210/endo-68-5-818

Kitay JI. 1964. Amelioration of cortisone-induced pituitary-adrenal suppression by estrad. *J Clin Endocrinol Metabol* **24:** 231–236. doi:10.1210/jcem-24-3-231

Klumbies E, Braeuer D, Hoyer J, Kirschbaum C. 2014. The reaction to social stress in social phobia: discordance between physiological and subjective parameters. *PLoS ONE* **9:** e105670. doi:10.1371/journal.pone.0105670

Kogler L, Gur RC, Derntl B. 2015. Sex differences in cognitive regulation of psychosocial achievement stress: brain and behavior. *Hum Brain Mapp* **36:** 1028–1042. doi:10.1002/hbm.22683

Kolber BJ, Muglia LJ. 2009. Defining brain region-specific glucocorticoid action during stress by conditional gene disruption in mice. *Brain Res* **1293:** 85–90. doi:10.1016/j.brainres.2009.03.061

Kret ME, De Gelder B. 2012. A review on sex differences in processing emotional signals. *Neuropsychologia* **50:** 1211–1221. doi:10.1016/j.neuropsychologia.2011.12.022

Kudielka BM, Kirschbaum C. 2005. Sex differences in HPA axis responses to stress: a review. *Biol Psychol* **69:** 113–132. doi:10.1016/j.biopsycho.2004.11.009

Kudwa AE, McGivern RF, Handa RJ. 2014. Estrogen receptor β and oxytocin interact to modulate anxiety-like behavior and neuroendocrine stress reactivity in adult male and female rats. *Physiol Behav* **129:** 287–296. doi:10.1016/j.physbeh.2014.03.004

Larkin JW, Binks SL, Li Y, Selvage D. 2010. The role of oestradiol in sexually dimorphic hypothalamic-pituitary-adrenal axis responses to intracerebroventricular ethanol administration in the rat. *J Neuroendocrinol* **22:** 24–32. doi:10.1111/j.1365-2826.2009.01934.x

Lund TD, Hinds LR, Handa RJ. 2006. The androgen 5α-dihydrotestosterone and its metabolite 5α-androstan-3β,17β-diol inhibit the hypothalamo-pituitary-adrenal response to stress by acting through estrogen receptor β-expressing neurons in the hypothalamus. *J Neurosci* **26:** 1448–1456. doi:10.1523/JNEUROSCI.3777-05.2006

Lupien SJ, Wilkinson CW, Brière S, Ménard C, Ng Ying Kin NM, Nair NP. 2002. The modulatory effects of corticosteroids on cognition: studies in young human populations. *Psychoneuroendocrinology* **27:** 401–416. doi:10.1016/S0306-4530(01)00061-0

Maeng LY, Milad MR. 2015. Sex differences in anxiety disorders: interactions between fear, stress, and gonadal hormones. *Horm Behav* **76:** 106–117. doi:10.1016/j.yhbeh.2015.04.002

Mason JW. 1975. A historical view of the stress field. *J Human Stress* **1:** 6–12. doi:10.1080/0097840X.1975.9940399

McEwen BS, Bowles NP, Gray JD, Hill MN, Hunter RG, Karatsoreos IN, Nasca C. 2015. Mechanisms of stress in the brain. *Nat Neurosci* **18:** 1353–1363. doi:10.1038/nn.4086

McGivern RF, Raum WJ, Salido E, Redei E. 1988. Lack of prenatal testosterone surge in fetal rats exposed to alcohol: alterations in testicular morphology and physiology. *Alcoholism: Clin Exp Res* **12:** 243–247.

McGlade E, Rogowska J, DiMuzio J, Bueler E, Sheth C, Legarreta M, Yurgelun-Todd D. 2020. Neurobiological evidence of sexual dimorphism in limbic circuitry of US Veterans. *J Affect Disord* **274**: 1091–1101. doi:10.1016/j .jad.2020.05.016

Nostro AD, Müller VI, Reid AT, Eickhoff SB. 2017. Correlations between personality and brain structure: a crucial role of gender. *Cereb Cortex* **27**: 3698–3712.

Ordaz S, Luna B. 2012. Sex differences in physiological reactivity to acute psychosocial stress in adolescence. *Psychoneuroendocrinology* **37**: 1135–1157. doi:10.1016/j .psyneuen.2012.01.002

Oyola M, Malysz A, Mani S, Handa R. 2016. Steroid hormone signaling pathways and sex differences in neuroendocrine and behavioral responses to stress. In *Sex differences in the central nervous system* (ed. RM Shansky), pp. 325–264. Elsevier, Philadelphia, PA.

Qian M, Zhang G, Luo S, Zhang S. 2000. Sex role inventory for college students (CSRI). *Chin J Psychol* **32**: 99–104.

Reschke-Hernández AE, Okerstrom KL, Bowles Edwards A, Tranel D. 2017. Sex and stress: men and women show different cortisol responses to psychological stress induced by the Trier social stress test and the Iowa singing social stress test. *J Neurosci Res* **95**: 106–114. doi:10.1002/ jnr.23851

Reul JM, de Kloet ER. 1985. Two receptor systems for corticosterone in rat brain: microdistribution and differential occupation. *Endocrinology* **117**: 2505–2511. doi:10.1210/ endo-117-6-2505

Rincón-Cortés M, Herman JP, Lupien S, Maguire J, Shansky RM. 2019. Stress: influence of sex, reproductive status and gender. *Neurobiol Stress* **10**: 100155. doi:10.1016/j.ynstr .2019.100155

Rivier C, Vale W. 1983. Modulation of stress-induced ACTH release by corticotropin-releasing factor, catecholamines and vasopressin. *Nature* **305**: 325–327. doi:10.1038/ 305325a0

Rood BD, de Vries GJ. 2011. Vasopressin innervation of the mouse (*Mus musculus*) brain and spinal cord. *J Comp Neurol* **519**: 2434–2474. doi:10.1002/cne.22635

Rosinger ZJ, Jacobskind JS, Bulanchuk N, Malone N, Fico D, Justice NJ, Zuloaga DG. 2019. Characterization and gonadal hormone regulation of a sexually dimorphic corticotropin-releasing factor receptor 1 cell group. *J Comp Neurol* **527**: 1056–1069. doi:10.1002/cne.24588

Rubinow DR, Schmidt PJ. 2019. Sex differences and the neurobiology of affective disorders. *Neuropsychopharmacology* **44**: 111–128. doi:10.1038/s41386-018-0148-z

Sapolsky RM, Romero LM, Munck AU. 2000. How do glucocorticoids influence stress responses? Integrating permissive, suppressive, stimulatory, and preparative actions. *Endocr Rev* **21**: 55–89.

Schienle A, Schäfer A, Stark R, Walter B, Vaitl D. 2005. Gender differences in the processing of disgust- and fear-inducing pictures: an fMRI study. *Neuroreport* **16**: 277–280. doi:10.1097/00001756-200502280-00015

Seale JV, Wood SA, Atkinson HC, Bate E, Lightman SL, Ingram CD, Jessop DS, Harbuz MS. 2004. Gonadectomy reverses the sexually diergic patterns of circadian and stress-induced hypothalamic-pituitary-adrenal axis activity in male and female rats. *J Neuroendocrinol* **16**: 516–524. doi:10.1111/j.1365-2826.2004.01195.x

Seale J, Wood S, Atkinson H, Harbuz M, Lightman SL. 2005. Postnatal masculinization alters the HPA axis phenotype in the adult female rat. *J Physiol* **563**: 265–274. doi:10 .1113/jphysiol.2004.078212

Selye H. 1937. Studies on adaptation. *J Endocrinol* **21**: 169–188. doi:10.1210/endo-21-2-169

Selye H. 1950. Stress and the general adaptation syndrome. *Br Med J* **1**: 1383–1392. doi:10.1136/bmj.1.4667.1383

Seo D, Ahluwalia A, Potenza MN, Sinha R. 2017. Gender differences in neural correlates of stress-induced anxiety. *J Neurosci Res* **95**: 115–125. doi:10.1002/jnr.23926

Sheng JA, Bales NJ, Myers SA, Bautista AI, Roueinfar M, Hale TM, Handa RJ. 2021a. The hypothalamic-pituitary-adrenal axis: development, programming actions of hormones, and maternal-fetal interactions. *Front Behav Neurosci* **14**: 601939. doi:10.3389/fnbeh.2020.601939

Sheng JA, Tan SM, Hale TM, Handa RJ. 2021b. Androgens and their role in regulating sex differences in the hypothalamic/pituitary/adrenal axis stress response and stress-related behaviors. *Androg Clin Res Ther* **2**: 261–274. doi:10.1089/andro.2021.0021

Shors TJ. 2016. A trip down memory lane about sex differences in the brain. *Philos Trans R Soc Lond B Biol Sci* **371**: 20150124. doi:10.1098/rstb.2015.0124

Smith CJ, Norman RL. 1987. Influence of the gonads on cortisol secretion in female rhesus macaques. *Endocrinology* **121**: 2192–2198. doi:10.1210/endo-121-6-2192

Solomon MB, Loftspring M, De Kloet AD, Ghosal S, Jankord R, Flak JN, Wulsin AC, Krause EG, Zhang R, Rice T, et al. 2015. Neuroendocrine function after hypothalamic depletion of glucocorticoid receptors in male and female mice. *Endocrinology* **156**: 2843–2853. doi:10.1210/en .2015-1276

Stephens MA, Mahon PB, McCaul ME, Wand GS. 2016. Hypothalamic-pituitary-adrenal axis response to acute psychosocial stress: effects of biological sex and circulating sex hormones. *Psychoneuroendocrinology* **66**: 47–55. doi:10.1016/j.psyneuen.2015.12.021

Stevens JS, Hamann S. 2012. Sex differences in brain activation to emotional stimuli: a meta-analysis of neuroimaging studies. *Neuropsychologia* **50**: 1578–1593. doi:10 .1016/j.neuropsychologia.2012.03.011

Stinnett GS, Westphal NJ, Seasholtz AF. 2015. Pituitary CRH-binding protein and stress in female mice. *Physiol Behav* **150**: 16–23. doi:10.1016/j.physbeh.2015.02.050

Stroud LR, Salovey P, Epel ES. 2002. Sex differences in stress responses: social rejection versus achievement stress. *Biol Psychiatry* **52**: 318–327. doi:10.1016/S0006-3223(02) 01333-1

Uhart M, Chong RY, Oswald L, Lin PI, Wand GS. 2006. Gender differences in hypothalamic-pituitary-adrenal (HPA) axis reactivity. *Psychoneuroendocrinology* **31**: 642–652. doi:10.1016/j.psyneuen.2006.02.003

Viau V, Meaney MJ. 1991. Variations in the hypothalamic-pituitary-adrenal response to stress during the estrous cycle in the rat. *Endocrinology* **129**: 2503–2511. doi:10 .1210/endo-129-5-2503

Viau V, Meaney MJ. 1996. The inhibitory effect of testosterone on hypothalamic-pituitary-adrenal responses to stress is mediated by the medial preoptic area. *J Neurosci*

16: 1866–1876. doi:10.1523/JNEUROSCI.16-05-01866 .1996

Viau V, Chu A, Soriano L, Dallman MF. 1999. Independent and overlapping effects of corticosterone and testosterone on corticotropin-releasing hormone and arginine vasopressin mRNA expression in the paraventricular nucleus of the hypothalamus and stress-induced adrenocorticotropic hormone release. *J Neurosci* 19: 6684–6693. doi:10 .1523/JNEUROSCI.19-15-06684.1999

Viau V, Bingham B, Davis J, Lee P, Wong M. 2005. Gender and puberty interact on the stress-induced activation of parvocellular neurosecretory neurons and corticotropin-releasing hormone messenger ribonucleic acid expression in the rat. *Endocrinology* 146: 137–146. doi:10.1210/en .2004-0846

Wang J, Korczykowski M, Rao H, Fan Y, Pluta J, Gur RC, McEwen BS, Detre JA. 2007. Gender difference in neural response to psychological stress. *Soc Cogn Affect Neurosci* 2: 227–239. doi:10.1093/scan/nsm018

Weiser MJ, Handa RJ. 2009. Estrogen impairs glucocorticoid dependent negative feedback on the hypothalamic-pituitary-adrenal axis via estrogen receptor α within the hypothalamus. *Neuroscience* 159: 883–895. doi:10.1016/j .neuroscience.2008.12.058

Williamson M, Viau V. 2008. Selective contributions of the medial preoptic nucleus to testosterone-dependent regulation of the paraventricular nucleus of the hypothalamus and the HPA axis. *Amer J Phys Reg* 295: 1020–1030.

Xin Y, Wu J, Yao Z, Guan Q, Aleman A, Luo Y. 2017. The relationship between personality and the response to acute psychological stress. *Sci Rep* 7: 16906. doi:10.1038/ s41598-017-17053-2

Young EA, Ribeiro SC, Ye W. 2007. Sex differences in ACTH pulsatility following metyrapone blockade in patients with major depression. *Psychoneuroendocrinology* 32: 503–507. doi:10.1016/j.psyneuen.2007.03.003

Yuan J, Li H, Long Q, Yang J, Lee TMC, Zhang D. 2021. Gender role, but not sex, shapes humans' susceptibility to emotion. *Neurosci Bull* 37: 201–216. doi:10.1007/s12264-020-00588-2

Yuen EY, Liu W, Karatsoreos IN, Feng J, McEwen BS, Yan Z. 2009. Acute stress enhances glutamatergic transmission in prefrontal cortex and facilitates working memory. *Proc Natl Acad Sci* 106: 14075–14079. doi:10.1073/pnas .0906791106

Zorn JV, Schür RR, Boks MP, Kahn RS, Joëls M, Vinkers CH. 2017. Cortisol stress reactivity across psychiatric disorders: a systematic review and meta-analysis. *Psychoneuroendocrinology* 77: 25–36. doi:10.1016/j.psyneuen.2016.11 .036

* Zucker I, Prendergast BJ, Beery AK. 2021. Pervasive neglect of sex differences in biomedical research. *Cold Spring Harb Perspect Biol* 14: a039156. doi: 10.1101/cshper spect.a039156.

Zuloaga DG, Morris JA, Jordan CL, Breedlove SM. 2008. Mice with the testicular feminization mutation demonstrate a role for androgen receptors in the regulation of anxiety-related behaviors and the hypothalamic-pituitary-adrenal axis. *Horm Behav* 54: 758–766. doi:10 .1016/j.yhbeh.2008.08.004

Zuloaga DG, Poort JE, Jordan CL, Breedlove SM. 2011. Male rats with the testicular feminization mutation of the androgen receptor display elevated anxiety-related behavior and corticosterone response to mild stress. *Horm Behav* 60: 380–388. doi:10.1016/j.yhbeh.2011.07.008

Zuloaga DG, Heck AL, De Guzman RM, Handa RJ. 2020. Roles for androgens in mediating the sex differences of neuroendocrine and behavioral stress responses. *Biol Sex Differ* 11: 44. doi:10.1186/s13293-020-00319-2

Sex Differences in Major Depressive Disorder (MDD) and Preclinical Animal Models for the Study of Depression

Elizabeth S. Williams, Michelle Mazei-Robison, and A.J. Robison

Department of Physiology, Michigan State University, East Lansing, Michigan 48824, USA
Correspondence: Robiso45@msu.edu

Depression and related mood disorders constitute an enormous burden on health, quality of life, and the global economy, and women have roughly twice the lifetime risk of men for experiencing depression. Here, we review sex differences in human brain physiology that may be connected to the increased susceptibility of women to major depressive disorder (MDD). Moreover, we summarize decades of preclinical research using animal models for the study of mood dysfunction that uncover some of the potential molecular, cellular, and circuit-level mechanisms that may underlie sex differences and disease etiology. We place particular emphasis on a series of recent studies demonstrating the central contribution of the circuit projecting from ventral hippocampus to nucleus accumbens and how inherent sex differences in the excitability of this circuit may predict and drive depression-related behaviors. The findings covered in this review underscore the continued need for studies using preclinical models and circuit-specific strategies for uncovering molecular and physiological mechanisms that could lead to potential sex-specific diagnosis, prognosis, prevention, and/or treatments for MDD and other mood disorders.

The World Health Organization recently named depression as the number one cause of disability globally (World Health Organization [WHO] 2017), highlighting its significant social and financial burden. Surprisingly, women have roughly twice the lifetime risk of men for experiencing depression (Kessler 2003), but the reasons for this remain unknown. Many genetic and environmental factors, such as inherited traits, early life adversity, and societal inequities, intersect at the level of the individual to produce depressive disorders. One of the most important of these factors is neurophysiology, which itself is influenced by countless external and internal forces, such as hormone status, stress exposure, age, and many others. The neurophysiology of depression is difficult to study in humans, as direct study of the structure and function of neurons and circuits requires invasive or terminal procedures. Accordingly, the field has relied on animal models for the study of depression, which often use stress and behavioral assays to assess depressive-like symptoms. Whereas previous work in animals has been instrumental in defining some basic physiological mechanisms important for depression, the bulk of research has been done in males and as such has only just begun to explore the etiology of sex differences in depression incidence. As our current body of knowledge regarding sex differences in depression is lacking, preclinical studies are needed to investigate neurophysiology, hormonal signaling, and other factors that intersect to produce these differences.

MAJOR DEPRESSIVE DISORDER (MDD) IN MEN AND WOMEN

One of the most striking statistics of major depressive disorder (MDD) is that women are twice as likely to develop the disorder as men (Seedat et al. 2009). Interestingly, when considering sex differences in MDD before puberty, boys are more likely to meet diagnostic criteria than girls (Douglas and Scott 2014). This begins to reverse at puberty, with female prevalence increasing to adulthood levels with each year of age post-puberty (Avenevoli et al. 2015). Depression symptom profiles also differ between men and women: women are more likely to report "atypical" symptoms of increased appetite and sleep, and to experience more fatigue and pain (Marcus et al. 2008).

Sex differences in depression rates may arise in part due to specific biological and environmental risk factors experienced by men and women (Kuehner 2017). Twin studies suggest that MDD carries ~37% heritability (Sullivan et al. 2000), and genome-wide association studies (GWAS) have uncovered a wide variety of potential heritable genetic loci that can contribute to depression pathogenesis (Flint and Kendler 2014). Hormonal influences can also affect individual susceptibility to MDD, with evidence supporting hormone fluctuation as one factor that increases MDD risk in women (Martel 2013). Differential activation of the HPA axis in men and women has also been implicated in the pathogenesis of MDD, as women experience hypoactivation of hypothalamic-pituitary-adrenal (HPA) responses that, evolutionarily, may protect a fetus from the effects of maternal stress (Kajantie and Phillips 2006). This hypoactivation, however, may deprive women of the possible protective effects of cortisol on depression-related changes to emotional brain circuitry

(Het et al. 2012). Many environmental factors such as gender-based violence and early life stress can also affect gender disparities in MDD. While not an exhaustive list, these biological and environmental risk factors, along with many others, underscore the complexity and heterogeneity of mood disorder etiology. Recognition of these factors, taken together with the neurophysiological bases for depression, inform our understanding of individual, sex- and gender-based variations in MDD experience.

PRECLINICAL ANIMAL MODELS OF DEPRESSION

Stress, along with other environmental factors, can be very influential on the highly plastic circuitry of the brain. Changes in activity or connectivity in the brain in response to an initial stressful scenario can alter many functional processes (e.g., learning, memory), resulting in an adaptation that better prepares the animal for future stressful events (McEwen 2008). However, when stress is chronic or traumatic, these changes can also coalesce to cause maladaptation to stress with accompanying susceptibility to depression or anxiety. Vulnerability to stress and subsequent maladaptation vary from individual to individual, with some individuals maintaining physiological and psychological function in the face of stress, while others suffer any number of pathologies in response to the same insult. Those that do not experience maladaptive stress response are considered resilient, and those that develop depression or other conditions are susceptible. Critically, differences in stress resilience can be modeled in rodents, and gene expression profiling of resilient and susceptible rodents has revealed that resilience may be an active mechanism (i.e., not simply the "absence" of susceptibility): more genes are differentially regulated in resilient mice compared to nonstressed controls than are regulated in susceptible mice (Krishnan et al. 2007). Two major stress paradigms for the study of depression, chronic social defeat stress (CSDS) and subchronic variable stress (SCVS), are discussed below.

Many life events (e.g., divorce, moving to a new city, illness) represent significant stressors and can precipitate the onset or recurrence of depression (Brown 1993). Life stressors in humans can lead to loss of social rank or loss of control over one's life circumstances (e.g., job loss leading to loss of income). To mimic human social stressors, rodent models have used social subordination as an ethologically relevant stressor that a rodent might face in its natural life. CSDS (Fig. 1, left) uses a resident-intruder paradigm in which the experimental male mouse (intruder) is placed into the home cage of an aggressive male mouse (resident), usually of a different strain (e.g., a smaller C57BL/6 mouse is placed into the home cage of a larger, sexually experienced CD-1 mouse; Fig. 1). Evaluation of susceptibility to CSDS is typically achieved through the social interaction (SI) test (Fig. 1, right), in which the experimental mouse is placed into an arena containing a novel aggressor mouse (social target) in a caged enclosure. Social withdrawal behavior is quantified by calculation of a social interaction ratio: the time spent in the interaction zone when the social

target is present over the time spent in the interaction zone when the social target is absent. The time spent in the corners by the experimental mouse is also indicative of social withdrawal, as susceptible mice are more likely to "hide" in the corners; resilient mice are eager to interact with the social target, as this interaction is rewarding.

In addition to exhibiting social withdrawal, susceptible mice also show other depression-like behaviors (Krishnan et al. 2007). Compared to nonstressed controls, susceptible mice show a significant decrease in body weight. They also demonstrate a reduction in sucrose preference, which is a measure of anhedonia (Willner et al. 1992). Both resilient and susceptible mice exhibit anxiety-like behavior, such as elevated corticosterone responses to forced swim test (FST) and decreased open arm time in the elevated plus maze (EPM).

The pharmacological validity of CSDS has been verified by experiments using standard antidepressant treatments (Tsankova et al. 2006;

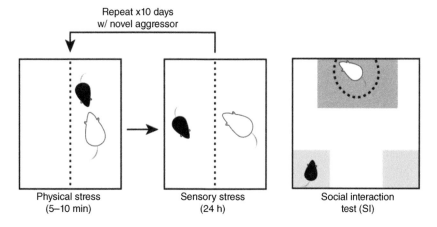

Repeat x10 days
w/ novel aggressor

Physical stress
(5–10 min)

Sensory stress
(24 h)

Social interaction
test (SI)

Figure 1. Chronic social defeat stress and social interaction test. Resident aggressor mice (e.g., white CD-1) are singly housed in cages for several days prior to testing to establish territory. The experimental intruder mouse (e.g., C57BL/6) is placed into the home cage of the aggressor and the two mice interact for 5–10 min (physical stress, *left*). Typically, the experimental mouse is quickly attacked by the aggressor mouse, and these attacks continue throughout the duration of the physical stress period. The experimental mouse is then moved to the other side of a perforated divider where it is safe from further attack, but can still see, hear, and smell the aggressor (sensory stress, *middle*). The experimental mouse is housed here for 24 h. The cycle then repeats for 10 or more days with the experimental mouse exposed to a novel aggressor mouse each day. Social interaction (SI) testing (*right*) is used following the chronic stress to evaluate social withdrawal as a measure of stress susceptibility. Susceptible mice exhibit social withdrawal and spend less time in the interaction zone (red) with the social target, and more time in the corners (yellow), withdrawn and isolated.

Zachariou et al. 2006). Treatment of susceptible animals with the antidepressants imipramine and fluoxetine reduced social withdrawal behavior caused by CSDS. Importantly, chronic administration was required, as an acute injection with these drugs did not reverse the withdrawal behavior (Zachariou et al. 2006). More recently, the fast-acting antidepressant ketamine has also been shown to reverse CSDS-induced phenotypes, and in this case a single injection was sufficient to reverse susceptibility (Donahue et al. 2014; Brachman et al. 2016).

SUBCHRONIC VARIABLE STRESS (SCVS)

One of the major limitations of CSDS is that it relies on territorial aggression between conspecific adult males and has not classically worked in females (Warren et al. 2020; although see Newman et al. 2019). In contrast, SCVS can be used with both sexes, with only female subjects subsequently susceptible to depression-related behaviors (Hodes et al. 2015; Williams et al. 2020). SCVS comprises 6 d of alternating stressors: foot shock, tail suspension, and restraint (Fig. 2, top). Following stress, mice are then assessed using a variety of assays: sucrose preference, SI, EPM, novelty-suppressed feeding (NSF) test, and splash test (Fig. 2, bottom). Following SCVS, female mice spend less time grooming in the splash test, display increased latency to eat in a novel environment, and have reduced sucrose preference (Hodes et al. 2015). Corticosterone is also only increased in stressed female mice, indicating dysregulation of the HPA axis (Hodes et al. 2015). These behavioral responses to SCVS do not occur in males, making SCVS useful for modeling the increased prevalence of MDD in women.

SEX DIFFERENCES IN STRESS RESPONSES

It is clear from the female-specific susceptibility in the SCVS paradigm that sex differences in stress response exist. These differences may account for the disproportionate number of women diagnosed with depression compared to men. Studies of stress responses in animal models of depression have revealed myriad sex differences, including differences in cellular and hormonal responses, brain circuits, synaptic and intrinsic plasticity, cognition, and emotional processing.

The HPA axis is a central component of stress responses and a key player in the pathophysiology of anxiety and depression. Corticotropin-releasing factor (CRF) initiates the HPA response. Neurons in the paraventricular nucleus of the hypothalamus (PVN) release CRF in response to stress, which reaches the pituitary and stimulates the release of adrenocorticotropic hormone (ACTH) into the systemic circulation, stimulating glucocorticoid release from the adrenal glands (Fig. 3), the PVN receives negative feedback regulation from the hippocampus (HPC) and other limbic structures (Herman et al. 2005). CRF also acts at the level of individual neurons via CRF receptors (Owens and Nemeroff 1991) and is released from the emotion-regulating limbic regions such as the bed nucleus of the stria terminalis (BNST) and central nucleus of the amygdala (CeA) (Walker et al. 2003) and can act throughout the brain as a neuromodulator. Indeed, CRF is elevated in cerebrospinal fluid of depressed humans (Nemeroff et al. 1984) and antidepressants decrease these levels (De Bellis et al. 1993; Heuser et al. 1998). CRF_1 receptor expression is also disturbed in many areas of the brain in depressed patients, including an increase in PVN (Wang et al. 2008) and possible compensatory decreases in cortical regions (Merali et al. 2004). Generally, HPA axis hormones (e.g., corticosterone) are elevated in female rodents compared to males at baseline (Kitay 1961). Female rodents, when stressed, release more ACTH and corticosterone than males and elevated levels also persist longer (Heinsbroek et al. 1991; Seale et al. 2004). These sex differences in rodents appear to reflect increased activation of the HPA system by CRF in females. PVN expression of CRF is elevated in females at baseline, particularly when estrogen levels are high (i.e., proestrus stage). Furthermore, stress increases CRF in female rodents only (Desbonnet et al. 2008; Iwasaki-Sekino et al. 2009). Epigenetic regulation of CRF is sexually dimorphic, with elevated methylation of *Crf* globally in the brain of female rats, but only in BNST and CeA in male rats (Sterrenburg et al. 2011). Particularly interesting is the differential expression of CRF_1 receptors in the CA1 region of the hip-

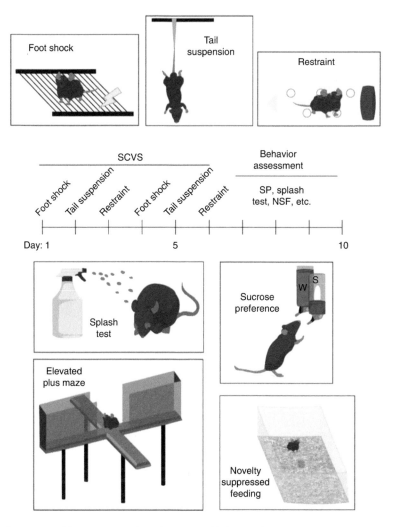

Figure 2. Subchronic variable stress (SCVS) and associated behavioral assays. SCVS comprises a 6-d battery of repeated stressors (*top*): foot shock, tail suspension, and restraint stress. Each stress session typically lasts 1 h per day. Following the stress period, a variety of behavioral assays are used to assess stress susceptibility (*bottom*): splash test, sucrose preference (SP), elevated plus maze (EPM), novelty suppressed feeding (NSF), and social interaction ([SI], not pictured). Splash test measures grooming behavior, with susceptible animals showing reduced grooming time when sprayed with a sticky solution. Sucrose preference indicates anhedonic response with two-bottle choice task, with susceptible animals showing no preference for sucrose drinking solution. EPM measures anxiety response, with "anxious" animals spending less time in open arms and more time in closed arms of the apparatus. Novelty suppressed feeding represents a more subtle measure of anxiety, with "anxious" animals exhibiting increased latency to eat (following overnight food deprivation) in a novel environment.

pocampus, an area of the brain central to emotional integration and regulation of behavioral responses to stress. Female rats, especially with elevated estrogen levels, have elevated CRF_1 receptor expression on pyramidal cells of the CA1 region (McAlinn et al. 2018); however, males have higher CRF_1 receptor expression on inhibitory interneurons in the dentate gyrus (DG). The same study demonstrated that CRF_1 receptor expression increases in CA1 pyramidal cells in male and female rodents following chronic immobilization stress, but male levels do not even reach the

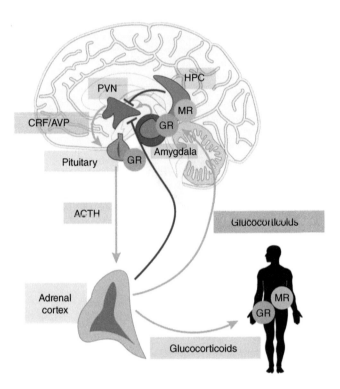

Figure 3. Hypothalamic-pituitary-adrenal (HPA) axis and hippocampal negative feedback regulation. The HPA axis is activated by stress and integrates stress responses. Anxiety responses via activation of the amygdala can exacerbate the stress response through its projections to the paraventricular nucleus of the hypothalamus (PVN). The hippocampus (HPC) is important for the evaluation of stress and is a site of glucocorticoid receptor (GR)- and mineralocorticoid (MR)-mediated negative feedback to the PVN. Release of the neuropeptides corticotropin-releasing hormone (CRH) and arginine vasopressin (AVP) from the hypothalamus promotes the release of adrenocorticotropin (ACTH) from the pituitary. ACTH then stimulates the release of glucocorticoids from the adrenal cortices. These hormones circulate to the body and brain and bind to intracellular steroid hormone receptors. (CRF) Corticotropin-releasing factor, (MR) mineralocorticoid receptor.

female baseline expression level. Collectively, these studies indicate that increased CRF signaling in females can differentially influence the HPA axis and central neuromodulation, contributing to observed sex differences in stress outcomes (Bangasser and Valentino 2012).

A variety of brain circuits are implicated in mediating sex differences in stress responses. The connection between the medial prefrontal cortex (mPFC) and amygdala is of particular interest, as the amygdala regulates responses to negative and traumatic stimuli (Hendler et al. 2003) and the mPFC regulates a wide range of functions including cognition and affect (Lieberman et al. 2019). Together, these brain regions play a significant role in behavioral responses to emotionally relevant contexts (Maren and Quirk 2004). Indeed,

mood disorders including MDD correlate with overactivity in the amygdala (Drevets 2003) as well as decreased mPFC activity (Pizzagalli et al. 2004). Moreover, chronic restraint stress in rodents causes restructuring and reorganization of dendrites in pyramidal cells of mPFC, which is posited to decrease the region's ability to suppress HPA activation (Radley et al. 2004). Chronic stress also causes hyperexcitability (Rosenkranz et al. 2010) and increased dendritic arborization (Vyas et al. 2002) in rat basolateral amygdala (BLA) neurons. Finally, BLA-projecting mPFC neurons increase in dendritic arborization in response to stress only in estrogen-treated, ovariectomized female rats (Shansky et al. 2010), suggesting that the response of this circuit to stress may be dependent on sex hormones.

Animal models for the study of depression have also highlighted sex differences in locus coeruleus (LC) norepinephrine (NE) circuitry (Bangasser and Valentino 2014). Dendritic architecture of LC neurons in female rats is more complex than in males, with female LC dendrites receiving significantly more synaptic input (Bangasser et al. 2011). Moreover, tonic and phasic firing patterns are linked to changes in arousal, and, while not significantly different at baseline, female LC neuron firing increases to a greater degree than male following swim stress in rats (Curtis et al. 2006). Taken together, these findings identify the LC NE circuitry as a key substrate in the mediation of sex differences in arousal, which may contribute to differences in male and female stress response and development of psychiatric disorders such as MDD. Together, data from humans and rodent models suggest widespread sex differences in stress-responsiveness throughout the brain that could contribute to the sex disparity in MDD. However, some of the most striking differences have been observed in the limbic structures, including HPC and nucleus accumbens (NAc).

HIPPOCAMPUS SEX DIFFERENCES IN STRESS AND DEPRESSION

The HPC is a central substrate in the pathophysiology of a wide range of mood disorders including MDD (Mervaala et al. 2000; Frodl et al. 2002; MacQueen et al. 2003), bipolar disorder (Beyer and Krishnan 2002), and posttraumatic stress disorder (PTSD) (Bremner et al. 1995). Many studies have identified varying degrees of hippocampal atrophy in patients with MDD (Sheline et al. 1996), with an inverse correlation between depression duration and hippocampus volume (Sheline et al. 1999). This atrophy is likely due to decreased neurogenesis in the DG (Malberg et al. 2000), as well as shrinking and retraction of dendrites and possible loss of glial support (Rajkowska 2000), and these stress-induced processes are often reversed by antidepressant treatment (Czeh et al. 2001). HPC abnormalities in MDD also extend to sex differences. For instance, women that do not respond to the antidepressant fluoxetine have smaller hippocampal volumes than those that do respond, an effect not observed in men (Vakili et al. 2000).

Preclinical models have revealed sex differences in female hippocampal function. CA1 dendritic spine density fluctuates with changes in estrogen levels: spine density decreases in late proestrus to estrus, then returns to early proestrus levels over a period of days (Woolley et al. 1990). These effects of estradiol on CA1 spine density are dependent upon glutamate neurotransmission via NMDA receptors (NMDARs), as density did not fluctuate when rats were treated with competitive NMDAR antagonists (Woolley and McEwen 1994). CA1 spine dynamics in response to stress are sexually dimorphic as male rat CA1 spine density increased following tail shock stress while female CA1 spine density decreased following the same stressor (Shors et al. 2001), and these opposite effects of stress on CA1 spine density are also NMDAR-dependent (Shors et al. 2004).

NUCLEUS ACCUMBENS SEX DIFFERENCES IN STRESS AND DEPRESSION

The NAc is a key substrate in the expression of depression-related behaviors in mouse models (Perrotti et al. 2004; Zachariou et al. 2006; Wilkinson et al. 2009; Vialou et al. 2010), and structural sex differences are apparent in the NAc. For example, distal dendritic spine density in NAc medium spiny neurons (MSNs) is higher in female rats (Forlano and Woolley 2010), indicating that excitatory synaptic connections in the NAc likely differ between the sexes. Indeed, there is sex-dependent variability in the morphology and distribution of dendritic spine synapses in NAc (Wissman et al. 2012).

Stress has sex-specific effects on glutamatergic inputs onto NAc (Brancato et al. 2017), which are implicated in the development of mood disorders (Russo and Nestler 2013; Thompson et al. 2015). Glutamatergic nerve terminals were reduced in the NAc of female but not male mice exposed to SCVS, but there were no post-stress sex differences in PSD95, nor sex differences in MSN spine density or morphology. This suggests sex differences in presynaptic glutamate neurotransmission in the NAc shell, supporting aberrations in excitatory signaling in this brain region as potential mediators of disparities in MDD prevalence in men and women. Excitatory neurotransmission to NAc comes from many regions, including amygdala, PFC, and ventral HPC (vHPC).

HPC To NAc CONNECTIVITY

The NAc is a key mediator of motivated behaviors and integrates its dopaminergic and glutamatergic inputs to drive motivated behaviors. Ventral tegmental area (VTA) dopaminergic input to the NAc mediates reinforcement of incentive salience (i.e., "wanting") of a reward (Berridge 2007; Day et al. 2007), while the glutamatergic input to NAc and midbrain dopamine neurons encodes information regarding reward-related environmental stimuli (Stuber et al. 2008). The excitatory input into the NAc allows integration of context and cues with goal-directed behavioral output (Pennartz et al. 2011). Dopamine signaling in the NAc modulates glutamatergic synaptic plasticity to integrate environmental information and prioritize the salience of some cues over others (Sun et al. 2008; Wolf and Ferrario 2010). Dysfunction of this integration leads to perturbations in reward processing that can result in aberrant motivated behavior. For example, an increase in the NAc's drive of motivated behavior may lead to addiction, and a weakening of this drive may lead to anhedonic behaviors such as those found in MDD and other mood disorders.

In rodents, three major glutamatergic projections to the NAc have been clearly demonstrated (Fig. 4): those from BLA, PFC, and vHPC (Britt et al. 2012). Moreover, functional magnetic resonance imaging (fMRI) has uncovered resting-state functional connectivity between the vHPC and NAc in humans (Kahn and Shohamy 2013). Projections from the vHPC are concentrated mainly at the medial NAc shell while projections from the BLA and PFC are mixed between the shell and the core (Britt et al. 2012). Furthermore, retrograde labeling studies indicate that the majority of all cells projecting to the medial shell are located in the vHPC, with fewer also coming from the BLA and PFC and the strength of vHPC inputs is strongest in NAc (Britt et al.

2012). The vHPC-NAc are uniquely able induce a stable depolarized state in NAc MSNs, increasing their likelihood of activity (O'Donnell and Grace 1995). This vHPC-mediated depolarization is necessary for the induction of spike firing by inputs from the PFC, suggesting that vHPC inputs are capable of "gating" the response of the NAc to other excitatory inputs (O'Donnell and Grace 1995).

Bagot et al. made crucial contributions to our understanding of the function of these projections with respect to stress outcomes, demonstrating opposite regulation of vHPC-NAc and mPFC-NAc activity by CSDS (Bagot et al. 2015). Using optogenetics to activate glutamatergic terminals in NAc arising from vHPC or mPFC to NAc, they found that glutamate release from vHPC-NAc projections is reduced in resilient mice, but oppositely, glutamate release from mPFC-NAc projections is increased in resilient mice. They then optically induced long-term depression (LTD) at vHPC-NAc synapses and found that weakening of vHPC-NAc connectivity resulted in an increase in social interaction to levels observed in resilient mice. Critically, LTD induced at mPFC-NAc inputs did not affect social interaction. Conversely, acute optical stimulation of vHPC-NAc projections reduced SI ratio following CSDS and increased immobility in the FST, thus promoting susceptibility to stress. The opposite was found in acute stimulation of mPFC-NAc projections, which increased SI ratio following CSDS in a pro-resilient fashion. This seminal study, while only examining male mice, highlights the importance of glutamatergic inputs to the NAc in regulating susceptibility to stress, and demonstrates that increasing vHPC-NAc activity is pro-depressive in male mice.

More recently, LeGates et al. (2018) demonstrated that vHPC-NAc synapses display activity-dependent plasticity and that their strength directly regulates reward behavior. Optical stimulation of vHPC-NAc synapses was sufficient to induce preference of a stimulation-paired chamber, and interruption of vHPC-NAc plasticity using optical inhibition was sufficient to abolish the preference of a chamber containing the natural reward of a social target. Furthermore, chronic multimodal stress (CMS) was sufficient

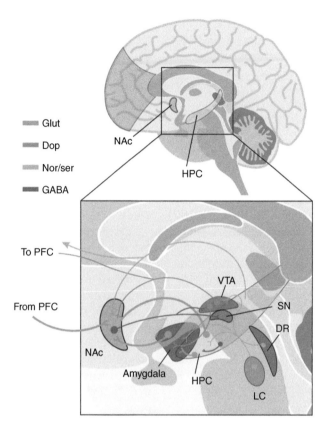

Glut
Dop
Nor/ser
GABA

Figure 4. Mood and reward circuitry. A simplified depiction of neural circuits that have been implicated in the pathophysiology of depression. The glutamatergic (green) projections to the nucleus accumbens (NAc) from hippocampus (HPC), amygdala, and prefrontal cortex (PFC) are prominently featured, and many subcortical regions that mediate reward, fear, and motivation are also depicted. This diagram also shows the innervation of many regions by monoaminergic projections from the ventral tegmental area ([VTA], blue dopaminergic projections to NAc, amygdala, PFC), as well as yellow noradrenergic projections from the locus coeruleus (LC) and serotonergic projections from the dorsal raphe (DR). The NAc also sends GABAergic (red) projections to the hypothalamus (not pictured), VTA, and substantia nigra (SN), and GABAergic interneurons are found within the HPC (and other structures). This diagram, while not exhaustive, highlights some of the major circuits involved in depression and referenced throughout this work.

to weaken vHPC synapses at D1R-MSNs, accompanied by a deficit in long-term potentiation (LTP) induction at these synapses, as well as an abolishment of the preference for a reward-paired chamber mentioned above. The link between stress-induced alterations at vHPC-NAc synapses and MDD was further strengthened by the mitigation of these alterations by the antidepressant fluoxetine. Chronic fluoxetine was sufficient to restore pre-stress sucrose preference levels, as well as the optically induced preference of the light stimulation-paired chamber. Furthermore,

chronic fluoxetine rescued the decrease in synaptic strength caused by chronic stress in D1R-MSNs and ameliorated the accompanying deficit in LTP induction. These findings highlight the key role of the connection between HPC and NAc in mediating reward and stress-related behaviors.

We recently demonstrated using whole-cell patch-clamp electrophysiology and retrograde labeling that vHPC-NAc neurons in female mice are more excitable than those neurons in male mice, and that this excitability difference correlated with the unique female susceptibility to SCVS discussed above (Williams et al. 2020). Furthermore, we found that orchiectomy of male mice elevated vHPC-NAc excitability and that this change was accompanied by an elevated susceptibility SCVS-induced reduction in sucrose preference. Female mice did not have a change in vHPC-NAc activity level when ovariectomized but, when exposed to chronic testosterone, did have reduction in excitability in this circuit and resilience to SCVS-induced reduction in sucrose preference. We were then able to show that the excitability of the circuit directly mediates the observed change in behavioral response to stress using designer receptors exclusively activated by designer drugs (DREADDs). Artificially increasing excitability of vHPC-NAc neurons in male mice caused susceptibility to SCVS, while using the same method to decrease excitability of vHPC-NAc neurons in female mice promoted resilience, as measured by sucrose preference. Most recently, Muir et al. showed that increased vHPC-NAc activity is predictive of anxiety-like behavior in the open field in both sexes and may predict some aspects of stress-induced decreases in social interaction, supporting the causal connection between vHPC-NAc activity and susceptibility to stress-induced behaviors associated with mood disorders (Muir et al. 2020).

All of these studies underscore the importance of the vHPC-NAc circuit as a critical component of behavioral response to stress, and identifies this circuit as uniquely suited to mediate sex-dependent differences in stress responses. Below, hormone signaling in the brain and its effects on stress and depression will be further explored.

HORMONE SIGNALING IN THE BRAIN— EFFECTS ON STRESS RESPONSE AND DEPRESSION

Estrogen signaling in the brain is primarily accomplished via intracellular estrogen receptors (ERs), which, when bound to estrogen, can translocate to the cell nucleus and bind to hormone response elements in target genes to affect their transcription; these events typically occur over hours (Woolley and Schwartzkroin 1998). Intracellular ERs are most densely expressed in the BNST, ventromedial nucleus of the hypothalamus, amygdala, various midbrain structures, and the pituitary (Pfaff and Keiner 1973), and more sparsely expressed in the HPC and cortex (Loy et al. 1988). Progesterone receptors (PRs) are generally expressed in the same pattern as ERs (Kato 1985). In situ hybridization and immunocytochemical studies suggest that most cells in the HPC that express intracellular ERs are interneurons (Pelletier et al. 1988; DonCarlos et al. 1991).

Progesterone, estrogen, and their derivatives can also have nongenomic, traditional cell-signaling effects on neuronal physiology, including on cellular excitability (Schumacher 1990). Progesterone metabolites allosterically bind $GABA_A$ receptors in the HPC, causing potentiated chloride conductance and dampened excitability (Majewska 1992). Excitatory postsynaptic potentials (EPSPs) in CA1 pyramidal cells of ovariectomized female rats are prolonged in rats that were "primed" with estrogen injections for 2 d prior to acute slice preparation (Wong and Moss 1992). This group also demonstrated that excitatory postsynaptic currents (EPSCs) were potentiated via the actions of second messenger cascades (e.g., cAMP signaling) (Gu and Moss 1996; Moss et al. 1997). These effects, mediated by cell signaling from hormone receptors on the cell membrane, make ovarian-derived hormones, especially estradiol, interesting potential mediators of sex differences in stress outcomes, particularly in the context of circuit-level excitability.

Estrogens also modulate synaptic plasticity in the hippocampus. LTP is enhanced during proestrus in rats (Warren et al. 1995), and LTP induction in CA1 is facilitated by estradiol treatment in ovariectomized rats (Córdoba Montoya and Carrer 1997). Furthermore, priming of ovariectomized rats with a combined treatment of estrogen and progesterone enhances voltage-gated calcium channel conductances in CA1 pyramidal neurons (Joels and Karst 1995). These studies complement the finding that estradiol has a potent effect on CA1 dendritic spine density (Gould et al. 1990), and, taken together, highlight important effects of estrogens on CA1 excitatory neurotransmission. Whether through transcriptional or cell-signaling mechanisms, ovarian hormones clearly play a role in regulating HPC physiology through modulation of excitability and synaptic transmission.

Estrous cycle may directly modulate behavioral responses to stress, including those dependent on the hippocampus. Male and female rodents react differently to stress in HPC-dependent learning tasks such as Morris water maze, with the most robust differences being identified when females are in proestrus (Shors and Leuner 2003). Furthermore, studies of ovariectomized rats have shown that stress enhances performance in the radial arm maze (another HPC-dependent measurement of spatial memory and learning) in animals with estrogen replacement, but not in those without (Bowman et al. 2002). Stress effects on CA1 dendritic spine density also appear to be estrogen-dependent: a single stressor can induce an increase in spine density in male rats but a decrease in female rats in proestrus (Shors et al. 2001), when spine density is usually the highest for females (Woolley et al. 1990).

Estrogen and stress can interact to cause sex differences in other regions as well. For example, the amygdala-dependent process of extinction of a conditioned fear response is impaired in estrogen-treated female rats (Toufexis et al. 2007). There is also evidence that estrogen can enhance stress effects in the mPFC, as proestrus or estrogen-replaced ovariectomized female rats demonstrated more deficiencies in mPFC-dependent working memory tasks (Shansky et al. 2004). Overall, estrogens have a clear effect on stress outcomes and brain function and are likely key mediators of sex differences in depression.

Testosterone has also been implicated in mediating stress outcomes and depression. In humans, hypogonadism leading to lower testosterone levels in men is associated with a higher prevalence of MDD (Shores et al. 2004; Zarrouf et al. 2009), and treatment with testosterone has been shown to alleviate depression symptoms in hypogonadal (Zarrouf et al. 2009) and eugonadal men (Kanayama et al. 2007; Walther et al. 2019). There is also evidence that low-dose testosterone treatment can improve depression scores and augment the effectiveness of tradi-

tional depression pharmacotherapies in treatment-resistant women (Miller et al. 2009).

Testosterone is metabolized to a variety of different effectors. For example, 5α-reductase converts testosterone into dihydrotestosterone (DHT), which is a more potent agonist at androgen receptors (ARs) and has a longer half-life than testosterone. DHT can then be converted by aldo-keto reductases to 3α-diol or 3β-diol. The former of these metabolites has a low affinity for AR but acts in the brain as an agonist at the $GABA_A$ receptor (Frye et al. 1996), and the latter acts primarily through ERβ in the brain to affect gene transcription (Pak et al. 2005). Like estrogens, androgens can also act through transcriptional and cell-signaling mechanisms. Androgens interact with intracellular receptors and diffuse into the nucleus to affect gene transcription via androgen receptor binding to DNA at hormone response elements (Nestler et al. 2009). Androgens and estrogens, as alluded to above, can also cause more rapid, nongenomic effects through activation of cell membrane-bound receptors (Cato et al. 2002). The rapid effects of androgens in the brain include modulation of neuronal excitability and plasticity. There is evidence that androgens acutely decrease neuronal excitability (Kubli-Garfias et al. 1982), but some studies indicate increased excitability via measurement of field potentials (Smith et al. 2002).

Testosterone has potent effects on anxiety- and depression-related behaviors in animal models. Orchiectomy of adult male rodents, for example, causes an increase in anxiety-like behaviors as assessed by marble burying, EPM and open field (OF) tests (Slob et al. 1981; Frye and Seliga 2001; Fernandez-Guasti and Martinez-Mota 2005), as well as behavioral despair (via FST) following CMS, which was accompanied by a decrease in hippocampal cell proliferation and neurogenesis (Wainwright et al. 2011). Testosterone relieves these anxiety like behaviors in intact adult male rodents (Bitran et al. 1993) as well as in the orchidectomized adult males of studies cited above. Particularly relevant to the current work, testosterone prevents anhedonia in aged adult male rats exposed to CMS if administered weeks before stress exposure (Herrera-Pérez et al.

2012). Interestingly, this effect was not observed if testosterone was administered following stress exposure, suggesting the necessity of testosterone before and during stress in male resilience. Further, prevention of anhedonia and behavioral despair by testosterone was likely mediated by aromatization to estradiol, as gonadectomized rats treated with DHT (insensitive to aromatase) still exhibited these depressive behaviors, but those treated with estradiol did not (Carrier and Kabbaj 2012a). However, other studies did implicate androgen metabolites in the mediation of stress resilience, as treatment with DHT or 3α-diol in intact adult male rodents decreases behavioral despair in FST (Frye and Walf 2009).

The effects of testosterone treatment in female rodents with respect to stress outcomes is even less clear. Testosterone, DHT, or 3α-diol treatment in intact female mice reduces anxiety-like behaviors as assessed by OF and EPM tests (Frye and Lacey 2001) and behavioral despair (FST) (Frye and Walf 2009). In contrast, however, other groups have observed no amelioration of anxiety (EPM) or depression (sucrose preference, NSF) behaviors following administration of testosterone in ovariectomized female rats (Carrier and Kabbaj 2012b). The difference between these observations may lie in the ovariectomy, as the studies demonstrating benefits of testosterone treatment in female animals did not include gonad removal. This could suggest that higher doses of testosterone are needed to induce anxiolytic or antidepressant effects in females, as maintaining the ovaries would accomplish this.

As alluded to above, the protective effects of androgens against stress-induced changes in behavior may in part be mediated through the HPC. Indeed, the volume of the hippocampal formation, its neuronal soma sizes, and extent of dendritic branching increase with androgen treatment in the perinatal period (Isgor and Sengelaub 1998, 2003). Additionally, testosterone enhances hippocampal neurogenesis induced by the antidepressant imipramine (Carrier and Kabbaj 2012b). The protective effects of neonatal androgens are associated with HPC neurogenesis and spine density, as male pups treated with the AR antagonist flutamide demonstrated depressive behaviors in FST and sucrose preference tests that correlated with decreased microtubule-associated protein-2 (MAP-2) labeling in the CA1 and DG and decreased spine density in CA1 (Zhang et al. 2010). Additionally, testosterone in orchidectomized adult male rodents may offer protection against oxidative damage and consequent changes in morphology (Meydan et al. 2010), and its metabolites in intact female adults appear to reduce HPC cell death induced by adrenalectomy (Frye and McCormick 2000). Testosterone and DHT also reverse orchidectomy-induced reduction of spine synapse density in males (independently of ER-mediated effects), while offering partial protection to females with contribution of the aromatization to estrogens (MacLusky et al. 2006). These morphological changes in response to steroid hormones may relate to HPC neuronal excitability. Teyler et al. (1980) observed field potential differences in HPC slices from rat brains in response to treatment with testosterone and estradiol: male rat slices showed increased excitability in response to estradiol but not testosterone, and female rat slices showed increased excitability in response to testosterone in diestrus but decreased excitability in proestrus.

CONCLUDING REMARKS

As described herein, genetics, neural circuits, hormones, and environment likely intersect at the level of the individual to shape behavioral responses to stress and dictate development of mood disorders. The past two decades of preclinical studies suggest that definition of the stress-responsive circuitry may guide future treatments for mood disorders such as depression and anxiety. To achieve this, we must elucidate regulatory mechanisms of these circuits, especially those mechanisms that differ between men and women. However, the mechanisms for differences in circuit physiology cannot yet be fully investigated in humans due to the invasiveness of existing methods; this underscores the continued need for studies using preclinical models and circuit-specific strategies for uncovering molecular and physiological mechanisms that could lead to potential treatments. The studies and findings summarized in this review begin to address the nature of circuit physiology in the context of stress and depression in male and female mice and may open the doors for the investigation of many new avenues for treatment or prevention of mood disorders.

ACKNOWLEDGMENTS

A.J.R. acknowledges support from the National Institutes of Mental Health (MH111604), the National Institutes of Neurological Disease and Stroke (NS085171), the National Institutes of Drug Abuse (DA040621), the National Institutes of Childhood Health and Disease (HD072968), and the Avielle Foundation. M.M.-R. acknowledges support from the National Institutes of Drug Abuse (DA039895).

REFERENCES

Avenevoli S, Swendsen J, He JP, Burstein M, Merikangas KR. 2015. Major depression in the national comorbidity survey-adolescent supplement: prevalence, correlates, and treatment. *J Am Acad Child Adolesc Psychiatry* 54: 37–44.e2. doi:10.1016/j.jaac.2014.10.010

Bagot RC, Parise EM, Peña CJ, Zhang HX, Maze I, Chaudhury D, Persaud B, Cachope R, Bolaños-Guzmán CA, Cheer JF, et al. 2015. Ventral hippocampal afferents to the nucleus accumbens regulate susceptibility to depression. *Nat Commun* 6: 7062. doi:10.1038/ncomms8062

Bangasser DA, Valentino RJ. 2012. Sex differences in molecular and cellular substrates of stress. *Cell Mol Neurobiol* 32: 709–723. doi:10.1007/s10571-012-9824-4

Bangasser DA, Valentino RJ. 2014. Sex differences in stress-related psychiatric disorders: neurobiological perspectives. *Front Neuroendocrinol* 35: 303–319. doi:10.1016/j.yfrne.2014.03.008

Bangasser DA, Zhang X, Garachh V, Hanhauser E, Valentino RJ. 2011. Sexual dimorphism in locus coeruleus dendritic morphology: a structural basis for sex differences in emotional arousal. *Physiol Behav* 103: 342–351. doi:10.1016/j.physbeh.2011.02.037

Berridge KC. 2007. The debate over dopamine's role in reward: the case for incentive salience. *Psychopharmacology (Berl)* 191: 391–431. doi:10.1007/s00213-006-0578-x

Beyer JL, Krishnan KR. 2002. Volumetric brain imaging findings in mood disorders. *Bipolar Disord* 4: 89–104. doi:10.1034/j.1399-5618.2002.01157.x

Bitran D, Kellogg CK, Hilvers RJ. 1993. Treatment with an anabolic-androgenic steroid affects anxiety-related behavior and alters the sensitivity of cortical GABAA receptors in the rat. *Horm Behav* 27: 568–583. doi:10.1006/hbeh.1993.1041

Bowman RE, Ferguson D, Luine VN. 2002. Effects of chronic restraint stress and estradiol on open field activity, spatial memory, and monoaminergic neurotransmitters in ovariectomized rats. *Neuroscience* 113: 401–410. doi:10.1016/S0306-4522(02)00156-2

Brachman RA, McGowan JC, Perusini JN, Lim SC, Pham TH, Faye C, Gardier AM, Mendez-David I, David DJ, Hen R, et al. 2016. Ketamine as a prophylactic against stress-induced depressive-like behavior. *Biol Psychiatry* 79: 776–786. doi:10.1016/j.biopsych.2015.04.022

Brancato A, Bregman D, Ahn HF, Pfau ML, Menard C, Cannizzaro C, Russo SJ, Hodes GE. 2017. Sub-chronic variable stress induces sex-specific effects on glutamatergic synapses in the nucleus accumbens. *Neuroscience* 350: 180–189. doi:10.1016/j.neuroscience.2017.03.014

Bremner JD, Randall P, Scott TM, Bronen RA, Seibyl JP, Southwick SM, Delaney RC, McCarthy G, Charney DS, Innis RB. 1995. MRI-based measurement of hippocampal volume in patients with combat-related posttraumatic stress disorder. *Am J Psychiatry* 152: 973–981. doi:10.1176/ajp.152.7.973

Britt JP, Benaliouad F, McDevitt RA, Stuber GD, Wise RA, Bonci A. 2012. Synaptic and behavioral profile of multiple glutamatergic inputs to the nucleus accumbens. *Neuron* 76: 790–803. doi:10.1016/j.neuron.2012.09.040

Brown GW. 1993. Stress: from synapse to syndrome. In *Life events and illness* (ed. Stanford S, Blanchard DC), pp. 20–40. Academic, Cambridge, MA.

Carrier N, Kabbaj M. 2012a. Extracellular signal-regulated kinase 2 signaling in the hippocampal dentate gyrus mediates the antidepressant effects of testosterone. *Biol Psychiatry* 71: 642–651. doi:10.1016/j.biopsych.2011.11.028

Carrier N, Kabbaj M. 2012b. Testosterone and imipramine have antidepressant effects in socially isolated male but not female rats. *Horm Behav* 61: 678–685. doi:10.1016/j.yhbeh.2012.03.001

Cato AC, Nestl A, Mink S. 2002. Rapid actions of steroid receptors in cellular signaling pathways. *Sci STKE* 2002: re9.

Córdoba Montoya DA, Carrer HF. 1997. Estrogen facilitates induction of long term potentiation in the hippocampus of awake rats. *Brain Res* 778: 430–438. doi:10.1016/S0006-8993(97)01206-7

Curtis AL, Bethea T, Valentino RJ. 2006. Sexually dimorphic responses of the brain norepinephrine system to stress and corticotropin-releasing factor. *Neuropsychopharmacology* 31: 544–554. doi:10.1038/sj.npp.1300875

Czeh B, Michaelis T, Watanabe T, Frahm J, de Biurrun G, van Kampen M, Bartolomucci A, Fuchs E. 2001. Stress-induced changes in cerebral metabolites, hippocampal volume, and cell proliferation are prevented by antidepressant treatment with tianeptine. *Proc Natl Acad Sci* 98: 12796–12801. doi:10.1073/pnas.211427898

Day JJ, Roitman MF, Wightman RM, Carelli RM. 2007. Associative learning mediates dynamic shifts in dopamine signaling in the nucleus accumbens. *Nat Neurosci* 10: 1020–1028. doi:10.1038/nn1923

De Bellis MD, Gold PW, Geracioti TD Jr, Listwak SJ, Kling MA. 1993. Association of fluoxetine treatment with reductions in CSF concentrations of corticotropin-releasing hormone and arginine vasopressin in patients with major depression. *Am J Psychiatry* 150: 656–657. doi:10.1176/ajp.150.4.656

Desbonnet L, Garrett L, Daly E, McDermott KW, Dinan TG. 2008. Sexually dimorphic effects of maternal separation stress on corticotrophin-releasing factor and vasopressin systems in the adult rat brain. *Int J Dev Neurosci* **26**: 259–268. doi:10.1016/j.ijdevneu.2008.02.004

Donahue RJ, Muschamp JW, Russo SJ, Nestler EJ, Carlezon WA Jr. 2014. Effects of striatal ΔFosB overexpression and ketamine on social defeat stress–induced anhedonia in mice. *Biol Psychiatry* **76**: 550-558. doi:10.1016/j.biopsych.2013.12.014

DonCarlos LL, Monroy E, Morrell JI. 1991. Distribution of estrogen receptor-immunoreactive cells in the forebrain of the female Guinea pig. *J Comp Neurol* **305**: 591–612. doi:10.1002/cne.903050406

Douglas J, Scott J. 2014. A systematic review of gender-specific rates of unipolar and bipolar disorders in community studies of pre-pubertal children. *Bipolar Disord* **16**: 5–15. doi:10.1111/bdi.12155

Drevets WC. 2003. Neuroimaging abnormalities in the amygdala in mood disorders. *Ann NY Acad Sci* **985**: 420–444. doi:10.1111/j.1749-6632.2003.tb07098.x

Fernandez-Guasti A, Martinez-Mota L. 2005. Anxiolytic-like actions of testosterone in the burying behavior test: role of androgen and GABA-benzodiazepine receptors. *Psychoneuroendocrinology* **30**: 762–770. doi:10.1016/j.psyneuen.2005.03.006

Flint J, Kendler KS. 2014. The genetics of major depression. *Neuron* **81**: 484–503. doi:10.1016/j.neuron.2014.01.027

Forlano PM, Woolley CS. 2010. Quantitative analysis of pre- and postsynaptic sex differences in the nucleus accumbens. *J Comp Neurol* **518**: 1330–1348.

Frodl T, Meisenzahl EM, Zetzsche T, Born C, Groll C, Jäger M, Leinsinger G, Bottlender R, Hahn K, Möller HJ. 2002. Hippocampal changes in patients with a first episode of major depression. *Am J Psychiatry* **159**: 1112–1118. doi:10.1176/appi.ajp.159.7.1112

Frye CA, Lacey EH. 2001. Posttraining androgens' enhancement of cognitive performance is temporally distinct from androgens' increases in affective behavior. *Cogn Affect Behav Neurosci* **1**: 172–182. doi:10.3758/CABN.1.2.172

Frye CA, McCormick CM. 2000. The neurosteroid, 3α-androstanediol, prevents inhibitory avoidance deficits and pyknotic cells in the granule layer of the dentate gyrus induced by adrenalectomy in rats. *Brain Res* **855**: 166–170. doi:10.1016/S0006-8993(99)02208-8

Frye CA, Seliga AM. 2001. Testosterone increases analgesia, anxiolysis, and cognitive performance of male rats. *Cogn Affect Behav Neurosci* **1**: 371–381. doi:10.3758/CABN.1.4.371

Frye CA, Walf AA. 2009. Depression-like behavior of aged male and female mice is ameliorated with administration of testosterone or its metabolites. *Physiol Behav* **97**: 266–269. doi:10.1016/j.physbeh.2009.02.022

Frye CA, Duncan JE, Basham M, Erskine MS. 1996. Behavioral effects of 3a-androstanediol II: hypothalamic and preoptic area actions via a GABAergic mechanism. *Behav Brain Res* **79**: 119–130. doi:10.1016/0166-4328(96)00005-8

Gould E, Woolley CS, Frankfurt M, McEwen BS. 1990. Gonadal steroids regulate dendritic spine density in hippocampal pyramidal cells in adulthood. *J Neurosci* **10**: 1286–1291. doi:10.1523/JNEUROSCI.10-04-01286.1990

Gu Q, Moss RL. 1996. 17β-estradiol potentiates kainate-induced currents via activation of the cAMP cascade. *J Neurosci* **16**: 3620–3629. doi:10.1523/JNEUROSCI.16-11-03620.1996

Heinsbroek RP, Van Haaren F, Feenstra MG, Endert E, Van de Poll NE. 1991. Sex- and time-dependent changes in neurochemical and hormonal variables induced by predictable and unpredictable footshock. *Physiol Behav* **49**: 1251–1256. doi:10.1016/0031-9384(91)90359-V

Hendler T, Rotshtein P, Yeshurun Y, Weizmann T, Kahn I, Ben-Bashat D, Malach R, Bleich A. 2003. Sensing the invisible: differential sensitivity of visual cortex and amygdala to traumatic context. *Neuroimage* **19**: 587–600. doi:10.1016/S1053-8119(03)00141-1

Herman JP, Ostrander MM, Mueller NK, Figueiredo H. 2005. Limbic system mechanisms of stress regulation: hypothalamo-pituitary-adrenocortical axis. *Prog Neuropsychopharmacol Biol Psychiatry* **29**: 1201–1213. doi:10.1016/j.pnpbp.2005.08.006

Herrera-Pérez JJ, Martínez-Mota L, Chavira R, Fernández-Guasti A. 2012. Testosterone prevents but not reverses anhedonia in middle-aged males and lacks an effect on stress vulnerability in young adults. *Horm Behav* **61**: 623–630. doi:10.1016/j.yhbeh.2012.02.015

Het S, Schoofs D, Rohleder N, Wolf OT. 2012. Stress-induced cortisol level elevations are associated with reduced negative affect after stress: indications for a mood-buffering cortisol effect. *Psychosom Med* **74**: 23–32. doi:10.1097/PSY.0b013e31823a4a25

Heuser I, Bissette G, Dettling M, Schweiger U, Gotthardt U, Schmider J, Lammers CH, Nemeroff CB, Holsboer F. 1998. Cerebrospinal fluid concentrations of corticotropin-releasing hormone, vasopressin, and somatostatin in depressed patients and healthy controls: response to amitriptyline treatment. *Depress Anxiety* **8**: 71–79. doi:10.1002/(SICI)1520-6394(1998)8:2<71::AID-DA5>3.0.CO;2-N

Hodes GE, Pfau ML, Purushothaman I, Ahn HF, Golden SA, Christoffel DJ, Magida J, Brancato A, Takahashi A, Flanigan ME, et al. 2015. Sex differences in nucleus accumbens transcriptome profiles associated with susceptibility versus resilience to subchronic variable stress. *J Neurosci* **35**: 16362–16376. doi:10.1523/JNEUROSCI.1392-15.2015

Isgor C, Sengelaub DR. 1998. Prenatal gonadal steroids affect adult spatial behavior, CA1 and CA3 pyramidal cell morphology in rats. *Horm Behav* **34**: 183–198. doi:10.1006/hbeh.1998.1477

Isgor C, Sengelaub DR. 2003. Effects of neonatal gonadal steroids on adult CA3 pyramidal neuron dendritic morphology and spatial memory in rats. *J Neurobiol* **55**: 179–190. doi:10.1002/neu.10200

Iwasaki-Sekino A, Mano-Otagiri A, Ohata H, Yamauchi N, Shibasaki T. 2009. Gender differences in corticotropin and corticosterone secretion and corticotropin-releasing factor mRNA expression in the paraventricular nucleus of the hypothalamus and the central nucleus of the amygdala in response to footshock stress or psychological stress in rats. *Psychoneuroendocrinology* **34**: 226–237. doi:10.1016/j.psyneuen.2008.09.003

Joels M, Karst H. 1995. Effects of estradiol and progesterone on voltage-gated calcium and potassium conductances in rat CA1 hippocampal neurons. *J Neurosci* **15:** 4289–4297. doi:10.1523/JNEUROSCI.15-06-04289.1995

Kahn I, Shohamy D. 2013. Intrinsic connectivity between the hippocampus, nucleus accumbens, and ventral tegmental area in humans. *Hippocampus* **23:** 187–192. doi:10.1002/hipo.22077

Kajantie E, Phillips DI. 2006. The effects of sex and hormonal status on the physiological response to acute psychosocial stress. *Psychoneuroendocrinology* **31:** 151–178. doi:10.1016/j.psyneuen.2005.07.002

Kanayama G, Amiaz R, Seidman S, Pope HG Jr. 2007. Testosterone supplementation for depressed men: current research and suggested treatment guidelines. *Exp Clin Psychopharmacol* **15:** 529-538. doi:10.1037/1064-1297.15.6.529

Kato J. 1985. Progesterone receptors in brain and hypophysis. In *Actions of progesterone on the brain current topics in neuroendocrinology* (ed. Ganten D, Pfaff D). Springer, Berlin.

Kessler RC. 2003. Epidemiology of women and depression. *J Affect Disord* **74:** 5–13. doi:10.1016/S0165-0327(02)00426-3

Kitay JI. 1961. Sex differences in adrenal cortical secretion in the rat. *Endocrinology* **68:** 818–824. doi:10.1210/endo-68-5-818

Krishnan V, Han MH, Graham DL, Berton O, Renthal W, Russo SJ, Laplant Q, Graham A, Lutter M, Lagace DC, et al. 2007. Molecular adaptations underlying susceptibility and resistance to social defeat in brain reward regions. *Cell* **131:** 391–404. doi:10.1016/j.cell.2007.09.018

Kubli-Garfias C, Canchola E, Arauz-Contreras J, Feria-Velasco A. 1982. Depressant effect of androgens on the cat brain electrical activity and its antagonism by ruthenium red. *Neuroscience* **7:** 2777–2782. doi:10.1016/0306-4522(82)90100-2

Kuehner K. 2017. Why is depression more common among women than among men? *The Lancet Psychiatry* **4:** 146–158. doi:10.1016/S2215-0366(16)30263-2

LeGates TA, Kvarta MD, Tooley JR, Francis TC, Lobo MK, Creed MC, Thompson SM. 2018. Reward behaviour is regulated by the strength of hippocampus–nucleus accumbens synapses. *Nature* **564:** 258–262. doi:10.1038/s41586-018-0740-8

Lieberman MD, Straccia MA, Meyer ML, Du M, Tan KM. 2019. Social, self, (situational), and affective processes in medial prefrontal cortex (MPFC): causal, multivariate, and reverse inference evidence. *Neurosci Biobehav Rev* **99:** 311–328. doi:10.1016/j.neubiorev.2018.12.021

Loy R, Gerlach JL, McEwen BS. 1988. Autoradiographic localization of estradiol-binding neurons in the rat hippocampal formation and entorhinal cortex. *Brain Res* **467:** 245–251. doi:10.1016/0165-3806(88)90028-4

MacLusky NJ, Hajszan T, Prange-Kiel J, Leranth C. 2006. Androgen modulation of hippocampal synaptic plasticity. *Neuroscience* **138:** 957–965. doi:10.1016/j.neuroscience.2005.12.054

MacQueen GM, Campbell S, McEwen BS, Macdonald K, Amano S, Joffe RT, Nahmias C, Young LT. 2003. Course of illness, hippocampal function, and hippocampal volume in major depression. *Proc Natl Acad Sci* **100:** 1387–1392. doi:10.1073/pnas.0337481100

Majewska MD. 1992. Neurosteroids: endogenous bimodal modulators of the GABAA receptor. Mechanism of action and physiological significance. *Prog Neurobiol* **38:** 379–395. doi:10.1016/0301-0082(92)90025-A

Malberg JE, Eisch AJ, Nestler EJ, Duman RS. 2000. Chronic antidepressant treatment increases neurogenesis in adult rat hippocampus. *J Neurosci* **20:** 9104–9110. doi:10.1523/JNEUROSCI.20-24-09104.2000

Marcus SM, Kerber KB, Rush AJ, Wisniewski SR, Nierenberg A, Balasubramani GK, Ritz L, Kornstein S, Young EA, Trivedi MH. 2008. Sex differences in depression symptoms in treatment-seeking adults: confirmatory analyses from the sequenced treatment alternatives to relieve depression study. *Compr Psychiatry* **49:** 238–246. doi:10.1016/j.comppsych.2007.06.012

Maren S, Quirk GJ. 2004. Neuronal signalling of fear memory. *Nat Rev Neurosci* **5:** 844–852. doi:10.1038/nrn1535

Martel MM. 2013. Sexual selection and sex differences in the prevalence of childhood externalizing and adolescent internalizing disorders. *Psychol Bull* **139:** 1221–1259. doi:10.1037/a0032247

McAlinn HR, Reich B, Contoreggi NH, Kamakura RP, Dyer AG, McEwen BS, Waters EM, Milner TA. 2018. Sex differences in the subcellular distribution of corticotropin-releasing factor receptor 1 in the rat hippocampus following chronic immobilization stress. *Neuroscience* **383:** 98–113. doi:10.1016/j.neuroscience.2018.05.007

McEwen BS. 2008. Central effects of stress hormones in health and disease: understanding the protective and damaging effects of stress and stress mediators. *Eur J Pharmacol* **583:** 174–185. doi:10.1016/j.ejphar.2007.11.071

Merali Z, Du L, Hrdina P, Palkovits M, Faludi G, Poulter MO, Anisman H. 2004. Dysregulation in the suicide brain: mRNA expression of corticotropin-releasing hormone receptors and GABA$_A$ receptor subunits in frontal cortical brain region. *J Neurosci* **24:** 1478–1485. doi:10.1523/JNEUROSCI.4734-03.2004

Mervaala E, Fohr J, Kononen M, Valkonen-Korhonen M, Vainio P, Partanen K, Partanen J, Tiihonen J, Viinamaki H, Karjalainen AK, et al. 2000. Quantitative MRI of the hippocampus and amygdala in severe depression. *Psychol Med* **30:** 117–125. doi:10.1017/S0033291799001567

Meydan S, Kus I, Tas U, Ogeturk M, Sancakdar E, Dabak DO, Zararsiz I, Sarsilmaz M. 2010. Effects of testosterone on orchiectomy-induced oxidative damage in the rat hippocampus. *J Chem Neuroanat* **40:** 281–285. doi:10.1016/j.jchemneu.2010.07.006

Miller KK, Perlis RH, Papakostas GI, Mischoulon D, Losifescu DV, Brick DJ, Fava M. 2009. Low-dose transdermal testosterone augmentation therapy improves depression severity in women. *CNS Spectr* **14:** 688–694. doi:10.1017/S1092852900023944

Moss RL, Gu Q, Wong M. 1997. Estrogen: nontranscriptional signaling pathway. *Recent Prog Horm Res* **52:** 33–68; discussion 68-39.

Muir J, Tse Y, Iyer E, Biris J, Cvetkovska V, Lopez J, Bagot R. 2020. Ventral-hippocampal afferents to nucleus accumbens encode both latent vulnerability and stress-induced susceptibility. *Biol Psychiatry* **88**: 843–854. doi:10.1016/j.biopsych.2020.05.021

Nemeroff CB, Widerlov E, Bissette G, Walleus H, Karlsson I, Eklund K, Kilts CD, Loosen PT, Vale W. 1984. Elevated concentrations of CSF corticotropin-releasing factor-like immunoreactivity in depressed patients. *Science* **226**: 1342–1344. doi:10.1126/science.6334362

Nestler EJ, Hyman SE, Malenka RC. 2009. *Molecular neuropharmacology: A foundation for clinical neuroscience.* McGraw-Hill Medical, New York.

Newman EL, Covington HE III, Suh J, Bicakci MB, Ressler KJ, DeBold JF, Miczek KA. 2019. Fighting females: neural and behavioral consequences of social defeat stress in female mice. *Biol Psychiatry* **86**: 657–668. doi:10.1016/j.biopsych.2019.05.005

O'Donnell P, Grace AA. 1995. Synaptic interactions among excitatory afferents to nucleus accumbens neurons: hippocampal gating of prefrontal cortical input. *J Neurosci* **15**: 3622–3639. doi:10.1523/JNEUROSCI.15-05-03622.1995

Owens MJ, Nemeroff CB. 1991. Physiology and pharmacology of corticotropin-releasing factor. *Pharmacol Rev* **43**: 425–473.

Pak TR, Chung WCJ, Lund TD, Hinds LR, Clay CM, Handa RJ. 2005. The androgen metabolite, 5α-androstane-3β, 17β-diol, is a potent modulator of estrogen receptor-β1-mediated gene transcription in neuronal cells. *Endocrinology* **146**: 147–155. doi:10.1210/en.2004-0871

Pelletier G, Liao N, Follea N, Govindan MV. 1988. Mapping of estrogen receptor-producing cells in the rat brain by in situ hybridization. *Neurosci Lett* **94**: 23–28. doi:10.1016/0304-3940(88)90264-9

Pennartz CM, Ito R, Verschure PF, Battaglia FP, Robbins TW. 2011. The hippocampal-striatal axis in learning, prediction and goal-directed behavior. *Trends Neurosci* **34**: 548–559. doi:10.1016/j.tins.2011.08.001

Perrotti LI, Hadeishi Y, Ulery PG, Barrot M, Monteggia L, Duman RS, Nestler EJ. 2004. Induction of ΔFosB in reward-related brain structures after chronic stress. *J Neurosci* **24**: 10594–10602. doi:10.1523/JNEUROSCI.2542-04.2004

Pfaff D, Keiner M. 1973. Atlas of estradiol-concentrating cells in the central nervous system of the female rat. *J Comp Neurol* **151**: 121–157. doi:10.1002/cne.901510204

Pizzagalli DA, Oakes TR, Fox AS, Chung MK, Larson CL, Abercrombie HC, Schaefer SM, Benca RM, Davidson RJ. 2004. Functional but not structural subgenual prefrontal cortex abnormalities in melancholia. *Mol Psychiatry* **9**: 393–405. doi:10.1038/sj.mp.4001469

Radley JJ, Sisti HM, Hao J, Rocher AB, McCall T, Hof PR, McEwen BS, Morrison JH. 2004. Chronic behavioral stress induces apical dendritic reorganization in pyramidal neurons of the medial prefrontal cortex. *Neuroscience* **125**: 1–6. doi:10.1016/j.neuroscience.2004.01.006

Rajkowska G. 2000. Postmortem studies in mood disorders indicate altered numbers of neurons and glial cells. *Biol Psychiatry* **48**: 766–777. doi:10.1016/S0006-3223(00)00950-1

Rosenkranz JA, Venheim ER, Padival M. 2010. Chronic stress causes amygdala hyperexcitability in rodents. *Biol Psychiatry* **67**: 1128–1136. doi:10.1016/j.biopsych.2010.02.008

Russo SJ, Nestler EJ. 2013. The brain reward circuitry in mood disorders. *Nat Rev Neurosci* **14**: 609–625. doi:10.1038/nrn3381

Schumacher M. 1990. Rapid membrane effects of steroid hormones: an emerging concept in neuroendocrinology. *Trends Neurosci* **13**: 359–362. doi:10.1016/0166-2236(90)90016-4

Seale JV, Wood SA, Atkinson HC, Harbuz MS, Lightman SL. 2004. Gonadal steroid replacement reverses gonadectomy-induced changes in the corticosterone pulse profile and stress-induced hypothalamic-pituitary-adrenal axis activity of male and female rats. *J Neuroendocrinol* **16**: 989–998. doi:10.1111/j.1365-2826.2004.01258.x

Seedat S, Scott KM, Angermeyer MC, Berglund P, Bromet EJ, Brugha TS, Demyttenaere K, de Girolamo G, Haro JM, Jin R, et al. 2009. Cross-national associations between gender and mental disorders in the World Health Organization World Mental Health Surveys. *Arch Gen Psychiat* **66**: 785–795. doi:10.1001/archgenpsychiatry.2009.36

Shansky RM, Glavis-Bloom C, Lerman D, McRae P, Benson C, Miller K, Cosand L, Horvath TL, Arnsten AF. 2004. Estrogen mediates sex differences in stress-induced prefrontal cortex dysfunction. *Mol Psychiatry* **9**: 531–538. doi:10.1038/sj.mp.4001435

Shansky RM, Hamo C, Hof PR, Lou W, McEwen BS, Morrison JH. 2010. Estrogen promotes stress sensitivity in a prefrontal cortex-amygdala pathway. *Cereb Cortex* **20**: 2560–2567. doi:10.1093/cercor/bhq003

Sheline YI, Wang PW, Gado MH, Csernansky JG, Vannier MW. 1996. Hippocampal atrophy in recurrent major depression. *Proc Natl Acad Sci* **93**: 3908–3913. doi:10.1073/pnas.93.9.3908

Sheline YI, Sanghavi M, Mintun MA, Gado MH. 1999. Depression duration but not age predicts hippocampal volume loss in medically healthy women with recurrent major depression. *J Neurosci* **19**: 5034–5043. doi:10.1523/JNEUROSCI.19-12-05034.1999

Shores MM, Sloan KL, Matsumoto AM, Moceri VM, Felker B, Kivlahan DR. 2004. Increased incidence of diagnosed depressive illness in hypogonadal older men. *Arch Gen Psychiatry* **61**: 162–167. doi:10.1001/archpsyc.61.2.162

Shors TJ, Leuner B. 2003. Estrogen-mediated effects on depression and memory formation in females. *J Affect Disord* **74**: 85–96. doi:10.1016/S0165-0327(02)00428-7

Shors TJ, Chua C, Falduto J. 2001. Sex differences and opposite effects of stress on dendritic spine density in the male versus female hippocampus. *J Neurosci* **21**: 6292–6297. doi:10.1523/JNEUROSCI.21-16-06292.2001

Shors TJ, Falduto J, Leuner B. 2004. The opposite effects of stress on dendritic spines in male vs. female rats are NMDA receptor-dependent. *Eur J Neurosci* **19**: 145–150. doi:10.1046/j.1460-9568.2003.03065.x

Slob AK, Bogers H, van Stolk MA. 1981. Effects of gonadectomy and exogenous gonadal steroids on sex differences in open field behaviour of adult rats. *Behav Brain Res* **2**: 347–362. doi:10.1016/0166-4328(81)90017-6

Smith MD, Jones LS, Wilson MA. 2002. Sex differences in hippocampal slice excitability: role of testosterone. *Neuroscience* **109**: 517–530. doi:10.1016/S0306-4522(01)00490-0

Cite this article as *Cold Spring Harb Perspect Biol* doi: 10.1101/cshperspect.a039198

Sterrenburg L, Gaszner B, Boerrigter J, Santbergen L, Bramini M, Elliott E, Chen A, Peeters BW, Roubos EW, Kozicz T. 2011. Chronic stress induces sex-specific alterations in methylation and expression of corticotropin-releasing factor gene in the rat. *PLoS ONE* **6:** e28128. doi:10.1371/journal.pone.0028128

Stuber GD, Klanker M, de Ridder B, Bowers MS, Joosten RN, Feenstra MG, Bonci A. 2008. Reward-predictive cues enhance excitatory synaptic strength onto midbrain dopamine neurons. *Science* **321:** 1690–1692. doi:10.1126/science.1160873

Sullivan PF, Neale MC, Kendler KS. 2000. Genetic epidemiology of major depression: review and meta-analysis. *Am J Psychiatry* **157:** 1552–1562. doi:10.1176/appi.ajp.157.10.1552

Sun X, Milovanovic M, Zhao Y, Wolf ME. 2008. Acute and chronic dopamine receptor stimulation modulates AMPA receptor trafficking in nucleus accumbens neurons cocultured with prefrontal cortex neurons. *J Neurosci* **28:** 4216–4230. doi:10.1523/JNEUROSCI.0258-08.2008

Teyler TJ, Vardaris RM, Lewis D, Rawitch AB. 1980. Gonadal steroids: effects on excitability of hippocampal pyramidal cells. *Science* **209:** 1017–1018. doi:10.1126/science.7190730

Thompson SM, Kallarackal AJ, Kvarta MD, Van Dyke AM, LeGates TA, Cai X. 2015. An excitatory synapse hypothesis of depression. *Trends Neurosci* **38:** 279–294. doi:10.1016/j.tins.2015.03.003

Toufexis DJ, Myers KM, Bowser ME, Davis M. 2007. Estrogen disrupts the inhibition of fear in female rats, possibly through the antagonistic effects of estrogen receptor α (ERα) and ERβ. *J Neurosci* **27:** 9729–9735. doi:10.1523/JNEUROSCI.2529-07.2007

Tsankova NM, Berton O, Renthal W, Kumar A, Neve RL, Nestler EJ. 2006. Sustained hippocampal chromatin regulation in a mouse model of depression and antidepressant action. *Nat Neurosci* **9:** 519–525. doi:10.1038/nn1659

Vakili K, Pillay SS, Lafer B, Fava M, Renshaw PF, Bonello-Cintron CM, Yurgelun-Todd DA. 2000. Hippocampal volume in primary unipolar major depression: a magnetic resonance imaging study. *Biol Psychiatry* **47:** 1087–1090. doi:10.1016/S0006-3223(99)00296-6

Vialou V, Robison AJ, Laplant QC, Covington HE III, Dietz DM, Ohnishi YN, Mouzon E, Rush AJ III, Watts EL, Wallace DL, et al. 2010. ΔFosB in brain reward circuits mediates resilience to stress and antidepressant responses. *Nat Neurosci* **13:** 745–752. doi:10.1038/nn.2551

Vyas A, Mitra R, Shankaranarayana Rao BS, Chattarji S. 2002. Chronic stress induces contrasting patterns of dendritic remodeling in hippocampal and amygdaloid neurons. *J Neurosci* **22:** 6810–6818. doi:10.1523/JNEUROSCI.22-15-06810.2002

Wainwright SR, Lieblich SE, Galea LA. 2011. Hypogonadism predisposes males to the development of behavioural and neuroplastic depressive phenotypes. *Psychoneuroendocrinology* **36:** 1327–1341. doi:10.1016/j.psyneuen.2011.03.004

Walker DL, Toufexis DJ, Davis M. 2003. Role of the bed nucleus of the stria terminalis versus the amygdala in fear, stress, and anxiety. *Eur J Pharmacol* **463:** 199–216. doi:10.1016/S0014-2999(03)01282-2

Walther A, Breidenstein J, Miller R. 2019. Association of testosterone treatment with alleviation of depressive symptoms in men: a systematic review and meta-analysis. *JAMA Psychiatry* **76:** 31–40. doi:10.1001/jamapsychiatry.2018.2734

Wang SS, Kamphuis W, Huitinga I, Zhou JN, Swaab DF. 2008. Gene expression analysis in the human hypothalamus in depression by laser microdissection and real-time PCR: the presence of multiple receptor imbalances. *Mol Psychiatry* **13:** 786–799, 741. doi:10.1038/mp.2008.38

Warren SG, Humphreys AG, Juraska JM, Greenough WT. 1995. LTP varies across the estrous cycle: enhanced synaptic plasticity in proestrus rats. *Brain Res* **703:** 26–30. doi:10.1016/0006-8993(95)01059-9

Warren BL, Mazei-Robison MS, Robison AJ, Iniguez SD. 2020. Can I get a witness? Using vicarious defeat stress to study mood-related illnesses in traditionally understudied populations. *Biol Psychiatry* **88:** 381–391. doi:10.1016/j.biopsych.2020.02.004

Wilkinson MB, Xiao G, Kumar A, LaPlant Q, Renthal W, Sikder D, Kodadek TJ, Nestler EJ. 2009. Imipramine treatment and resiliency exhibit similar chromatin regulation in the mouse nucleus accumbens in depression models. *J Neurosci* **29:** 7820–7832. doi:10.1523/JNEUROSCI.0932-09.2009

Williams ES, Manning CE, Eagle AL, Swift-Gallant A, Duque-Wilckens N, Chinnusamy S, Moeser A, Jordan C, Leinninger G, Robison AJ. 2020. Androgen-dependent excitability of mouse ventral hippocampal afferents to nucleus accumbens underlies sex-specific susceptibility to stress. *Biol Psychiatry* **87:** 492–501. doi:10.1016/j.biopsych.2019.08.006

Willner P, Muscat R, Papp M. 1992. Chronic mild stress-induced anhedonia: a realistic animal model of depression. *Neurosci Biobehav Rev* **16:** 525–534. doi:10.1016/S0149-7634(05)80194-0

Wissman AM, May RM, Woolley CS. 2012. Ultrastructural analysis of sex differences in nucleus accumbens synaptic connectivity. *Brain Struct Funct* **217:** 181–190. doi:10.1007/s00429-011-0353-6

Wolf ME, Ferrario CR. 2010. AMPA receptor plasticity in the nucleus accumbens after repeated exposure to cocaine. *Neurosci Biobehav Rev* **35:** 185–211. doi:10.1016/j.neubiorev.2010.01.013

Wong M, Moss RL. 1992. Long-term and short-term electrophysiological effects of estrogen on the synaptic properties of hippocampal CA1 neurons. *J Neurosci* **12:** 3217–3225. doi:10.1523/JNEUROSCI.12-08-03217.1992

Woolley CS, McEwen BS. 1994. Estradiol regulates hippocampal dendritic spine density via an *N*-methyl-D-aspartate receptor-dependent mechanism. *J Neurosci* **14:** 7680–7687. doi:10.1523/JNEUROSCI.14-12-07680.1994

Woolley CS, Schwartzkroin PA. 1998. Hormonal effects on the brain. *Epilepsia* **39:** S2–S8. doi:10.1111/j.1528-1157.1998.tb02601.x

Woolley CS, Gould E, Frankfurt M, McEwen BS. 1990. Naturally occurring fluctuation in dendritic spine density on adult hippocampal pyramidal neurons. *J Neurosci*

10: 4035–4039. doi:10.1523/JNEUROSCI.10-12-04035 .1990

World Health Organization (WHO). 2017. *Depression and other common mental disorders: global health estimates.* World Health Organization, Geneva, Switzerland.

Zachariou V, Bolanos CA, Selley DE, Theobald D, Cassidy MP, Kelz MB, Shaw-Lutchman T, Berton O, Sim-Selley LJ, Dileone RJ, et al. 2006. An essential role for ΔFosB in the nucleus accumbens in morphine action. *Nat Neurosci* **9:** 205–211. doi:10.1038/nn1636

Zarrouf FA, Artz S, Griffith J, Sirbu C, Kommor M. 2009. Testosterone and depression: systematic review and meta-analysis. *J Psychiatr Pract* **15:** 289–305. doi:10.1097/01 .pra.0000358315.88931.fc

Zhang JM, Tonelli L, Regenold WT, McCarthy MM. 2010. Effects of neonatal flutamide treatment on hippocampal neurogenesis and synaptogenesis correlate with depression-like behaviors in preadolescent male rats. *Neuroscience* **169:** 544–554. doi:10.1016/j.neuroscience.2010.03 .029

Sex Differences in Mast Cell–Associated Disorders: A Life Span Perspective

Emily Mackey[1,2] and Adam J. Moeser[1]

[1]Department of Large Animal Clinical Sciences, College of Veterinary Medicine, Michigan State University, East Lansing, Michigan 48864, USA

[2]Comparative Biomedical Sciences Program, North Carolina State University, College of Veterinary Medicine, Raleigh, North Carolina 27603, USA

Correspondence: moeserad@msu.edu

Mast cells are critical innate immune effectors located throughout the body that are crucial for host defense mechanisms via orchestrating immune responses to a variety of host and environmental stimuli necessary for survival. The role of mast cells in brain development and behavior, meningeal function, and stress-related disorders has also been increasingly recognized. While critical for survival and development, excessive mast cell activation has been linked with an increasing number of inflammatory, stress-associated, and neuroimmune disorders including allergy/anaphylaxis, autoimmune diseases, migraine headache, and chronic pain disorders. Further, a strong sex bias exists for mast cell–associated diseases with females often at increased risk. Here we review sex differences in human mast cell–associated diseases and animal models, and the underlying biological mechanisms driving these sex differences, which include adult gonadal sex hormones as well the emerging organizational role of perinatal gonadal hormones on mast cell activity and development.

M ast cells are hematopoietic-derived innate immune cells ubiquitously located in the body and are potent orchestrators and effector cells in the immune response. Moreover, many mast cell–associated disorders including irritable bowel syndrome, migraine, chronic pain, allergy/anaphylaxis, and autoimmune disease, exhibit a strong sex bias in which females are more susceptible. Adult sex hormones may explain some of the causes of sex-biased disease responses; however, this idea is challenged by the fact that sex biases in many mast cell–associated disorders are evident in prepubertal children. In this review, we will discuss sex biases in mast cell disease susceptibility with a focus on factors of sex that may be at play. We highlight the diverse roles of mast cells in health and disease in context with the interaction of sex. Further, we call attention to early life as a critical period in shaping mast cell function and discuss early life environmental and host interactions that may influence mast cell disease susceptibility across the life span. The findings in this review provide insights for therapeutic targets for mast cell disorders with a potential for identifying sex-specific therapies in pediatric and adult disease.

MAST CELL BIOLOGICAL ROLES IN HEALTH AND DISEASE

Mast cells are important effector cells of the immune system and play critical roles in inflammatory disease. Mast cells arise from hematopoietic $CD34^+/CD117^+$ stem cells that circulate as committed progenitors in blood, and populate all tissues, especially at host–environment interfaces such as the skin, the lung, and the gut (Abraham and Malaviya 1997). Mast cells are long-lived, able to survive months or years, and can proliferate in tissues in response to certain stimuli (Abraham and Malaviya 1997; Galli et al. 2005). The most distinguishable characteristic of mast cells are the 50- to 200-electron dense granules that occupy the majority of the cytoplasm of mature mast cells (Krystel-Whittemore et al. 2015). Stored within these granules are large amounts of preformed proinflammatory mediators (e.g., histamine, serotonin, proteases, etc.), which when released initiate rapid and robust physiologic effects on the vasculature, epithelium, and nervous systems (Krystel-Whittemore et al. 2015).

The strategic location of mast cells next to vessels and nerves and at host–environment interfaces, as well as the plethora of immune mediators they contain and synthesize, positions them to be involved in a variety of pathologic conditions.

Mast cells are best known for their association with pathologic conditions such as allergy and anaphylaxis where aberrant mast cell activation leads to damage of host tissues. However, mast cells are indispensable to the host and no humans lacking mast cells have ever been described (Wong et al. 2014). Further, mast cell–like cells have been described in an early ancestor of vertebrates, *Ciona intestinalis*, pointing to an ancient origin of mast cells (>500 million years ago), well before the development of adaptive immunity and therefore IgE-mediated allergy (Wong et al. 2014). The fact that mast cells have persisted throughout vertebrate evolution reinforces their importance in immune responses against infectious diseases, including those by parasites, bacteria, and viruses.

Mast cells are positioned at host–environment interfaces with a repertoire of receptors ready to react to pathogen-associated molecular patterns and other signals of infection (Gilfillan et al. 2009). Further, mast cells have a kinetic advantage over other sentinel immune cells through their ability to release a multitude of preformed immune mediators instantaneously at a site of infection. Mast cells promote clearance of bacteria and prevent dissemination of infections in the peritoneum, bladder, lung, gut, and skin (Malaviya et al. 1996; Wei et al. 2005; Siebenhaar et al. 2007; Sutherland et al. 2008; Shelburne et al. 2009). Mast cells release mediators in response to parasite infection, including histamine to increase vascular permeability, and smooth muscle contraction to expel parasites, and proteases that are directly toxic to parasites (Vermillion et al. 1988; McKean and Pritchard 1989; McDermott et al. 2003). During viral infections, mast cells are involved in recruitment of immune cells, but further questions remain with regard to other functions of mast cells during viral infections (Orinska et al. 2005).

EPIDEMIOLOGY OF SEX DIFFERENCES IN MAST CELL–ASSOCIATED DISEASE

An increasing number of clinical and epidemiologic studies demonstrate sex biases in mast cell–associated disorders, often with females exhibiting increased prevalence and severity of disease. Diseases classically associated with mast cells including allergy, anaphylaxis, and asthma (Webb and Lieberman 2006; Osman et al. 2007a; Poulos et al. 2007; Kool et al. 2016; Acker et al. 2017). Further, evidence has mounted supporting important roles of mast cells in other immune-related disorders. Clinical and preclinical animal research have linked mast cells with autoimmune diseases such as multiple sclerosis (MS) (Orton et al. 2006) and rheumatoid arthritis (Alamanos and Drosos 2005), which also exhibit a female sex bias, occurring in women at 4 times the rate of men (Chiaroni-Clarke et al. 2016). Heightened mast cell activation near sensory neurons has been linked with symptom severity of chronic pain disorders such as irritable bowel syndrome (Sperber et al. 2017), migraine (Stovner et al. 2007), interstitial cystitis (Berry et al. 2011), and fibromyalgia (Walitt et al. 2015), which all occur much more frequently in women (Mogil 2012). Further, mast cells degranulation has been implicated in sex-specific pain disorders in women such as vulvodynia and endometriosis (Bornstein et al. 2004; Anaf et al. 2006).

Research into the mechanisms underlying sex differences in mast cell–mediated disease has largely focused on adult gonadal sex hormones as the main drivers of sex differences. This focus is likely due in part to the sex reversal in prevalence of allergic rhinitis, food allergy, and asthma from a male predominance prior to puberty to a female predominance in adulthood (Kelly and Gangur 2009; Fröhlich et al. 2017; Pinart et al. 2017; Hohmann et al. 2019). However, childhood prevalence of a number of mast cell–associated diseases still exhibit a female predominance such as eczema (Ballardini et al. 2013), autoimmune disease (Chiaroni-Clarke et al. 2016), irritable bowel syndrome (Korterink et al. 2015), migraine (Victor et al. 2010), fibromyalgia (Kashikar-Zuck and Ting 2014), interstitial cystitis (Vaz et al. 2012), and general chronic pain (King et al. 2011). With regard to asthma, sex-specific childhood prevalence rates of asthma have shifted between 1989 and 2004 moving from male-biased prevalence to no sex difference, while eczema and allergic rhinitis prevalence changed from a male bias to a female bias (Osman et al. 2007b). Similarly, the prevalence of childhood food allergies, which have previously been reported to be more common in males, have now shifted toward a female predominance (Branum and Lukacs 2008; McGowan and Keet 2013). This shift toward a female bias in asthma and atopy (heightened immune responses to common skin, airway, and food allergens) could be attributed to several factors including enhanced sensitivity in the evaluation of clinical symptoms and diagnosis of atopy, or a decrease in the bias of health professionals to more readily diagnose boys with atopy (Osman et al. 2007b). Together, sex differences in susceptibility to mast cell–associated disease exist throughout life, and therefore early life and adult sex-based factors in pathogenesis of these disorders are likely involved and covered later in this review.

ROLE OF MAST CELLS IN SEX DIFFERENCES IN IgE-MEDIATED ALLERGY/ANAPHYLAXIS AND EXPERIMENTAL AUTOIMMUNE ENCEPHALOMYELITIS

As described above, a female sex bias is evident in mast cell–associated disorders, but the specific contribution of mast cells to this sex bias is poorly understood. Here, we focus on studies that have demonstrated sex differences in mast cells activation in animal models of passive systemic anaphylaxis and experimental autoimmune encephalomyelitis (EAE), and how sex differences in mast cells may drive sex-biased mast cell disease risk and severity.

Sex Differences in IgE-Mediated Passive Systemic Anaphylaxis and the Role of Mast Cells

As discussed above, a distinguishing feature of mast cells is the capacity to store and release preformed granule mediators (e.g., histamine, serotonin, proteases), which allows them to respond rapidly to threats in the microenvironment. However, aberrant or excessive activation of mast cells, such as in allergy and anaphylaxis, can have potentially life-threatening consequences. Allergy is the most recognized negative consequence of mast cell activation and afflicts 30%–40% of people worldwide (Pawankar et al. 2012). Mast cells are the main effector cell in IgE-mediated allergic reactions. Infusing IgE antibodies into wild-type mice, mast cell–deficient mice, and mast cell knockin mice demonstrates that many of the allergen-induced biologic responses are entirely dependent upon mast cell degranulation, including cardiopulmonary collapse in anaphylaxis, tissue swelling, fibrin deposition, airway hyperreactivity, and immune cell recruitment (Galli et al. 2005). Allergies develop when innocuous environmental antigens elicit type 2 immune responses through activation of CD4[+] T helper type 2 cells that engage B cells to produce allergen-specific IgE antibodies (Hofmann and Abraham 2009). IgE binds to FcεRI receptors on tissue mast cells and basophils, which mainly circulate in low numbers in the blood (Voehringer 2013). Mast cells are activated upon reexposure to the allergen, which cross-links IgE antibodies bound to FcɛRI receptors, causing degranulation and release of multiple preformed mediators (e.g., histamine, serotonin, proteases, etc.), referred to as an immediate hypersensitivity reaction (Hofmann and Abraham 2009). These mediators are responsible for increasing vascular permeability, smooth muscle contraction, and mucus secretion as well as signaling nerves and recruiting other immune cells, which contribute to allergic symptoms like redness, swelling, itching, runny nose, and diarrhea (Hofmann and Abraham 2009). In allergic asthma, mast cell mediators cause bronchoconstriction, mucus secretion, and respiratory mucosal edema leading to reduced air flow and wheeze (Hofmann and Abraham 2009). Other allergic disorders including allergic rhinitis, eczema, urticaria, and food allergy also manifest through IgE-mediated mast cell activation (Galli and Tsai 2012). Further, systemic IgE-mediated mast cell activation can result in anaphylaxis, a catastrophic immune response that can rapidly lead to death if untreated (Galli and Tsai 2012).

In line with the female sex bias in humans, IgE-mediated induction of clinical scores of anaphylaxes and pathophysiologic outcomes (e.g., blood histamine levels, tissue mast cell activation, airway hyperreactivity, and hypothermia) in mouse models are more severe in female animals (Mackey et al. 2016, 2020; Hox et al. 2015). Despite several studies reporting heightened severity of mast cell–mediated anaphylaxis in females, the factors responsible for these sex differences have been conflicting. Ovariectomy reduced allergic airway inflammation in female rats (Ligeiro de Oliveira et al. 2004) and the hypothermia and airway inflammation in C57Bl6 mice (Hox et al. 2015), while estrogen replacement in ovariectomized mice reversed the effects of ovariectomy. Further, Hox et al. (2015) reported that females express higher levels of endothelial nitric oxide synthase (eNOS) and phosphorylated eNOS, compared with males and ovariectomized females, and that administration of the NOS inhibitor, L-NAME, reduced hypothermia responses in female mice but not males. Overall, these finding imply that gonadal estrogens enhance the severity of IgE-mediated hypothermia and airway inflammation via en-hanced eNOS, thus contributing to heightened anaphylactic responses in females. In contrast, we found that ovariectomy of adult C57Bl6 females and castration of males had no significant effect on IgE-mediated histamine release or hypothermia (Mackey et al. 2020). Instead, Mackey et al. (2020) concluded that these sex differences were a result of organizational effects of perinatal androgens, rather than adult androgens, shaping sex differences in mast cell histamine release and hypothermia, as discussed in more detail below.

Role of Mast Cells in Autoimmune Disease and Their Contribution to Sex Differences in Disease Severity

Autoimmune diseases are a result of the immune system failing to recognize self from non-self molecules, resulting in activation of self-reactive lymphocytes by innate immune cells to respond to self-antigens (Brown and Hatfield 2012). As mentioned previously, mast cells are associated with several autoimmune diseases in humans, including Sjogren's syndrome (Konttinen et al. 2000), chronic idiopathic urticaria (Saini et al. 2009), experimental vasculitis (Ishii et al. 2009), rheumatoid and idiopathic arthritis (Nigrovic et al. 2007; Sullivan 2007), bullous pemphigoid (Wintroub et al. 1978), and systemic lupus erythematosus (Kaczmarczyk-Sekula et al. 2015). But the mast cell's role in MS, a chronic inflammatory disorder of the central nervous system (CNS) characterized by mononuclear cell infiltration of the brain and demyelination of neurons, has received the most attention. Mast cells are observed within the demyelinated plaques in the brains of humans with MS and elevated levels of tryptase are found in the cerebrospinal fluid of MS patients (Krüger et al. 1990; Rozniecki et al. 1995). In the rodent model of MS, EAE, mast cell–deficient mice (Wsh mice) exhibited less severe EAE-type lesions, including less severe demyelination (Brown et al. 2002; Desbiens et al. 2016). Perivascular mast cells are a source of vasoactive and proinflammatory mediators that increase the permeability of the blood–brain barrier and can recruit immune cells to the brain (Sayed et al. 2010), which is important for MS pathogenesis. Further, administration of a mast cell stabilizer drug reduced severity of EAE (Brown and Hatfield 2012), supporting a critical role of mast cells in the MS-associated pathology and in general autoimmunity as reviewed previously (Brown and Hatfield 2012).

The biological mechanisms underlying the female sex bias in autoimmune diseases has received considerable attention. Sex differences in autoimmune disease are thought to be mediated through various mechanisms, including sex hormone and genetic factors and a complex interplay between innate and adaptive immune cells. Mast cells have been proposed as potentially critical players mediating sex differences in autoimmune diseases. It was also demonstrated that IL-33 is a critical protective factor in the EAE model, through a mechanism involving IL-33-mediated activation of ILC2 cells and generation of a Th2 response to confer protection against inflammation-induced CNS lesions in males. In vitro experiments with bone marrow–derived mast cells (BMMCs) showed that testosterone, through binding to androgen receptors (ARs) expressed on mast cells, induces IL-33 production and release from male, but not female mast cells, which the authors speculated could be a mast cell–dependent mechanism driving sex differences in the EAE model. The authors also showed that cultured female mast cells release greater amounts of inflammatory IL-1β and TNF-α, which was speculated as a potential driver of a Th-17-dominated anti-myelin response contributing to heightened disease severity in females (Russi et al. 2018). Together, these studies support a critical role of mast cells in the pathogenesis of EAE lesions and that mechanism involving sex-dependent effects of testosterone on male mast cells could protect males from autoimmune-associated CNS lesions.

ADDRESSING SEX BIASES IN MAST CELL–ASSOCIATED DISEASES IN CHILDREN: ROLE OF PERINATAL ANDROGENS

While much attention has been directed toward adult gonadal sex hormones as a mechanism mediating sex differences in mast cell–associated disease, epidemiological evidence demonstrates these same sex biases in prepubertal children, as noted above. Therefore, sex-based factors independent of adult gonadal sex hormones are likely playing a significant role. A defining mechanism of sexual differentiation in the fetus and newborn is the perinatal androgen surge in males. During this organizational period, testes secrete gonadal androgens (testosterone and dihydrotestosterone) in both prenatal and early postnatal life, which is known to induce permanent masculinizing effects on many body systems. Testosterone can also be converted to estradiol via the aromatase enzyme to drive defeminizing effects via estrogen receptors. Androgens and estrogens with their respective receptors bind to hormone-responsive elements on target DNA sequences, which attracts cofactors that form transcriptional complexes to regulate gene transcription (Hiort 2013). Many coregulators have inherent histone acetyltransferase and methyltransferase activity, which can alter the epigenetic state of the genome and change gene expression (Arnold et al. 2012). These epigenetic marks to histones or DNA can have long-lasting impacts on gene expression, permanently influencing the function and phenotype of cells (Arnold et al. 2012). In contrast to males, sexual differentiation in females during the organizational period occurs in the absence of the androgen surge and is thought to be largely independent of gonadal hormones as the ovaries are relatively quiescent until puberty. The actions of the perinatal testicular androgen surge on many tissues during this critical period organize many body systems to respond to pubertal gonadal hormones or the "activational" phase in adulthood. In other words, the perinatal androgen surge leads to a fixed program of gene expression in target cells throughout life, which controls the function and phenotype of that cell, including future responsiveness to hormones.

Perinatal Androgens in Males as Major Drivers of Sex Differences in IgE-Mediated Anaphylaxis through Alterations in Mast Cell Development

As mentioned above, studies by Mackey et al. (2016, 2020) demonstrated that female mice exhibit greater concentrations of serum histamine levels, which coincided with more severe hypothermia, compared with male mice. Similar sex differences were also observed in mice exposed to psychological restraint stress (Mackey et al. 2016; D'Costa et al. 2019). Moreover, sex differences in IgE-mediated passive systemic anaphylaxis were also demonstrated in 14-day-old mice, which suggested that factors other than adult gonadal sex hormones could play a significant role in the sex differences. Administration of testosterone propionate (TP) to pregnant dams resulted in "masculinized" mast cell–mediated responses, with reduced blood histamine concentrations and reduced severity of hypothermia response in female offspring at neonatal and adult stages. On the other hand, inhibition of the perinatal androgens surge in males by administration of the anti-androgen di-(2-ethylhexyl) phthalate (DEHP) to pregnant dams, resulted in heightened severity or "feminized" IgE-mediated anaphylaxis responses in male offspring. Together, these studies established the perinatal androgen surge as a critical event driving long-term attenuated mast cell responses in males. This work further introduced the concept that host, genetic, and environmental-associated alterations in androgens during the critical perinatal period can have lasting developmental consequences on mast cell–associated disease severity.

Developmental Effects of Perinatal Androgens on Mast Cell Granule Mediator Storage and Its Role in Sex Differences in Mast Cell–Mediated Anaphylaxis

Studies by Mackey et al. (2016, 2020) showed that heightened anaphylaxis in female mice was associated with higher levels of blood histamine. Examination of tissue mast cells revealed no significant differences in mast cell number or degranulation between the sexes. Isolation of peritoneal mast cells from female and male mice and rats demonstrated that female mast cells contain higher levels of preformed granule-associated mediators (e.g., histamine, serotonin, tryptase, etc.). Electron microscopy also revealed more electron-dense granules in female mast cells compared with male mast cells, further indicating that

females have an increased capacity to store granule mediators. Furthermore, infusion of female mast cells into mast cell–deficient male mice recipients, and vice versa, showed that sex differences in IgE-mediated passive systemic anaphylaxis were determined by the sex of the donor mast cells and not the recipient. Additionally, Mackey et al. (2020) showed that mast cells derived in vitro from bone marrow hematopoietic stem cell precursors (BMMCs) obtained from female, male, and perinatally androgenized females exhibit sex differences in mast cell granule mediator content, with male and androgenized female-derived BMMCs exhibiting significantly reduced levels of histamine, serotonin, and proteases compared with female BMMCs. Together, these findings demonstrate that biological sex and high perinatal androgen levels alter the developmental programming of mast cells toward reduced granule mediators' storage and release, contributing to reduced clinical severity of mast cell–mediated anaphylaxis.

Perinatal Androgens as Modulators of Early Brain Mast Cell Activity, Brain Development, and Sex-Related Behavior

While best known for their pathophysiological role in inflammatory diseases, mast cells have been implicated in normal development of organ systems, specifically the nervous system. Mast cells are abundant in the brain and meninges and have been shown to modulate neuronal physiology and behavior. With regard to early developmental programming, an interplay between sex hormone levels during the perinatal organizational period, mast cells and sexual dimorphism of the preoptic area (POA), and sex-specific behaviors has been demonstrated. In a study by Lenz et al. (2018), male rat fetuses exhibited significantly greater numbers of mast cells in the POA compared with females. Administration of masculinizing doses of estradiol (metabolite of testosterone via aromatase enzyme) during the perinatal period increased POA mast cell numbers and degranulation in females to the level of males, and was associated with "male-typical" synaptic patterning in the brain and masculinization of adult sex behavior (Lenz et al. 2018).

Similarly, activation of fetal brain mast cells by allergen challenge in pregnant dams resulted in sex-specific changes in dendritic spine patterning of POA neurons, and lifelong alterations in adult behavior, such that female offspring exhibited male-typical sexual behavior, while male offspring exhibited decreased male-typical behaviors (Lenz et al. 2019). Together, these studies highlighted not only the important developmental role of mast cells in brain development and adult sexual behavior but demonstrated that perinatal estrogen organizes sex differences in POA neurons and lasting developmental changes in adult sexual behavior via modulation of mast cell number and activation.

IMPACT OF ESTROUS CYCLE AND SEX HORMONE LEVELS ON TISSUE MAST CELL NUMBERS

Tissue mast cell numbers can change dramatically throughout the estrous cycle and correspond with changing concentrations of sex hormones. In rodents, the density of mast cells fluctuates with the estrous cycle in the dura mater, mammary gland, ovary, and uterus, but not in the jejunum or colon (Aydın et al. 1998; Bradesi et al. 2001; Boes and Levy 2012; Jing et al. 2012). Similarly, mast cell numbers in the ovary and uterus change during the estrous cycle in dogs, cats, cows, and goats, with highest numbers of mast cells in reproductive organs during the follicular phase when estrogen levels are high (Özen et al. 2007; Karaca et al. 2008; Hamouzova et al. 2017, 2020). Exogenous estrogen treatment of rodents increased the numbers of mast cells in the dura mater, uterus, and mammary gland, but not in the jejunum or colon (Bradesi et al. 2001; Jensen et al. 2010; Boes and Levy 2012; Jing et al. 2012). In contrast to mast cell numbers, mast cell degranulation was not influenced by the estrous cycle or exogenous estrogen in the dura mater; however, mast cell degranulation was reduced in the ovary and uterus during estrus, when estrogen levels were high (Aydın et al. 1998; Boes and Levy 2012). Ovariectomy reduced mast cell degranulation in response to substance P in the jejunum, but increased mast cell degranulation in response to substance P in the colon (Bradesi et al. 2001).

In addition, G protein estrogen receptor agonist treatment in rats induced mast cell degranulation in colonic tissues (Xu et al. 2020). Taken together, estrogen increases mast cell numbers in a tissue-specific manner during fetal development, adulthood, and during the estrous cycle, but does not appear to have clear role in mast cell degranulation. The mechanisms by which estrogen modulates mast cell number remains poorly understood and represents a critical gap in knowledge that has implications for development and mast cell–associated disease, particularly those in which disease severity or clinical flares are more common during specific stages of the estrous cycle. For example, estrogen and progesterone have been suggested to enhance mast cell–associated disease in females. Further, symptoms of asthma and irritable bowel syndrome fluctuate with the menstrual cycle, suggesting involvement of estrogen and progesterone (Vrieze et al. 2003; Meleine and Matricon 2014). Visceral pain sensitivity to colorectal distension increased in rats during estrus and proestrus, when estrogen levels are high (Moloney et al. 2016). While collectively these studies present an interesting association between sex hormone levels, mast cell number, and disease severity, a definitive mechanistic link between these factors remains unknown.

SEX HORMONE INTERACTIONS WITH MAST CELLS

The above studies demonstrate a link between sex hormones levels and mast cell number and activity; however, precisely how sex hormones influence mast cell activity remains to be elucidated. Several studies in human and rodent mast cells have shown that mast cells express receptors for estrogen, progesterone, and testosterone (AR) (Pang et al. 1995; Zaitsu et al. 2007; Jensen et al. 2010), which implies that sex hormones could be acting directly on mast cells to alter their function. In support of this, several in vitro mast cell experiments have examined the effects of sex hormones on mast cell activation. Exogenous estrogen administration induces relatively minor degranulation responses (~5%) across several mast cell lines and sources including RBL-2H3 rodent mast cell line, HMC-1 human mast cell line, primary BMMCs, and rat peritoneal mast cells (Narita et al. 2007; Zaitsu et al. 2007; Zhu et al. 2018). Mast cell degranulation responses to estrogens was shown to be through a membrane-associated form of estrogen receptor α, and not the nuclear receptor (Zaitsu et al. 2007). However, a more recent study showed no effect of exogenous estrogen on mast cell degranulation in HMC-1 cells (Chen et al. 2010), but estrogen reduced the secretion of TNF-α, IL-6, and IL-1β after ionophore stimulus (Kim et al. 2001). High doses of progesterone also induce a minor mast cell degranulation response in HMC-1 cells (Jensen et al. 2010). In contrast, progesterone inhibited degranulation of rat peritoneal mast cells stimulated with substance P (Vasiadi et al. 2006). Interestingly, low doses of estradiol and progesterone increased histamine release (∼5%) in female rat peritoneal mast cells but had no effect on male rat peritoneal mast cells (Munoz-Cruz et al. 2015). Together, these studies suggest that estrogen and progesterone have modest effects on in vitro mast cell degranulation responses.

The influence of testosterone on mast cell activity has been evaluated to a lesser extent. However, a recent meta-analysis analyzing immune function outcomes in experimental models after manipulation of sex hormones (122 studies) found testosterone had a moderate immunosuppressive effect on immune function, while estrogen had no effect on immune function overall (Foo et al. 2017). In line with this conclusion, animal models of parasitism found orchidectomy increases clearance of helminths, while ovariectomy had no effect (Tiuria et al. 1994). However, testosterone administration in female rodents decreased the ability to expel intestinal parasites (Tiuria et al. 1995). Conversely, testosterone has been shown to be protective in models of asthma and autoimmune disease. Castrated male mice exhibited enhanced IL-33-mediated lung inflammation, which was attributed to the effects of testosterone in reducing the innate lymphoid cell 2 (ILC2) populations (Laffont et al. 2017), but the specific role of mast cells in this study was not evaluated. Human and rodent mast cells express ARs (Chen et al. 2010; Russi et al. 2018; Mackey et al. 2020). Of note, BMMCs derived

from female mice have higher AR (*Ar*) gene expression compared to male-derived BMMCs (Mackey et al. 2016); however, Russi et al. 2018 did not detect sex differences in IL-33 protein expression in BMMCs via flow cytometry. Several studies have shown that response to testosterone and other mast cell stimuli may be sex dependent. Low doses of testosterone and dihydrotestosterone induced degranulation in female, but not male, rat peritoneal mast cells (Chen et al. 2010; Munoz-Cruz et al. 2015). As mentioned previously, testosterone induced IL-33 gene and protein expression in male but not female BMMCs (Russi et al. 2018). Overall, these studies show that both male and female mast cells express functional ARs, but testosterone effects on mast cell gene and protein expression can be sex specific.

INFLUENCE OF THE SEX CHROMOSOME COMPLEMENT ON THE MAST CELL

Sex hormones have been shown to be a major player in determining sex differences in body plan and function, but sex chromosome complement also plays an important role in sex-specific development of the body. Mammalian sex determination begins in the zygote where an inherent imbalance of genes encoded by heteromorphic sex chromosomes (XX vs. XY) influences gonadal development. Most notably, in the male zygote, the Y-linked gene *Sry* induces a series of cellular and molecular events that lead to the formation of testes. In the absence of *Sry*, ovaries develop in the XX female through an active, albeit less understood series of events. Sex determination of the gonads based on sex chromosomes leads to lifelong sex differences in the secretion of sex steroid hormones.

Beyond the role of sex chromosome complement on gonadal differentiation and steroid hormone production, there is potential for direct sex chromosome effects driving sex differences in mast cells. Females inherit two X chromosomes, whereas males carry one X chromosome and one Y chromosome. To avoid duplicate gene expression in females, one X chromosome is randomly silenced during X chromosome inactivation in females. Previous studies showed up to 7% of

Cite this article as *Cold Spring Harb Perspect Biol* doi: 10.1101/cshperspect.a039172

genes escape silencing from the inactive X chromosome in mouse tissues (Berletch et al. 2015) and 15% in human cells (Carrel and Willard 2005). Considering the X chromosome contains approximately 1100 genes, of which many are involved in immunity, the effect of gene escape from X chromosome inactivation in mast cells may have consequences for the development and activity of the cells (Libert et al. 2010). Further, gene expression from the Y chromosome in male cells may create variances that impact cell phenotype and function (Cortez et al. 2014). The complement of sex chromosomes in a cell can also influence genome-wide DNA methylation in a hormone-independent fashion (Wijchers and Festenstein 2011). Mast cells derived from adult male and female mice exhibit markedly different transcriptomes (Mackey et al. 2016), but to our knowledge, the direct effects of the sex chromosome complement in mast cells has not been elucidated. Exploration of sex differences in fetal-derived mast cells prior to the perinatal androgen surge would provide insight into the effects of the sex chromosome complement. Moreover, use of the "four core genotypes" mouse model, which allows for dissociation of sex hormones and sex chromosomes, would be valuable to determine possible sex chromosome complement effects on mast cell phenotype and function (Arnold and Chen 2009).

EVOLUTIONARY PERSPECTIVE OF MAST CELL SEX DIFFERENCES: DO FEMALES OR MALES HOLD THE ADVANTAGE?

As mentioned above, mast cells are critical effector cells of the innate immune system with a primary role in protection from infection through their ability to rapidly orchestrate innate and adaptive immune responses. In general, males are more susceptible to pathogenic infections and exhibit higher mortality rates than females (vom Steeg and Klein 2016). For example, males have increased prevalence of viral infections from HIV, influenza, and hepatitis B and bacterial infections including tuberculosis, legionellosis, and campylobacter (vom Steeg and Klein 2016). Sex differences exist in the SARS-CoV-2 pandemic with men developing more severe manifestations

of disease (M:F relative ratio of hospitalization 1.5:1), which results in higher death rates (M:F relative ratio of case fatalities 1.7:1) compared with women (Gebhard et al. 2020). In parallel, antibody responses to bacterial and viral vaccines, including influenza and hepatitis B, are higher in females than males (Klein and Flanagan 2016). A comprehensive review of parasitic infections in relation to sex demonstrated that 46 out of 53 (86.8%) parasite species (protozoa, nematodes, trematodes, and cestodes) had higher infection rates in males for a variety of species (Klein 2004). Sex differences in infection rates and vaccine responses have been largely demonstrated in adults, but this trend is also evident in children, as newborn male children are more vulnerable to infections and are more likely to die than female children (Sawyer 2012). Further, boys under the age of 5 have higher rates of protozoan, trematode, and nematode infections than girls (Flanagan et al. 2015). Female children before the onset of puberty also have greater antibody responses to many vaccines (hepatitis B, diphtheria, pertussis, pneumococcal, rabies, measles, and RTS,S against malaria) (Klein and Flanagan 2016). While the specific contributions of mast cells to sex differences in infectious disease remains to be elucidated, the well-established role of mast cells in host defense, pathogen clearance, and host survival suggests that heightened mast responses in females may confer an important survival advantage. However, in conditions where pathogen disease pressures are lower, such as in developed countries, the heightened mast cell responses in females may underlie the increased risk for chronic mast cell–associated diseases.

CONCLUDING REMARKS

Biological sex can be a protective factor against disease, sometimes to a greater extent than offered by drugs or therapies. Discovering the sex-biasing factors that can protect from disease may lead to the development of novel targets of therapy. Here, we discussed the sometimes complex and multifactorial nature of sex differences in mast cells and how this might underlie the sex bias in mast cell–associated diseases throughout life (Fig. 1). While adult sex hormones clearly

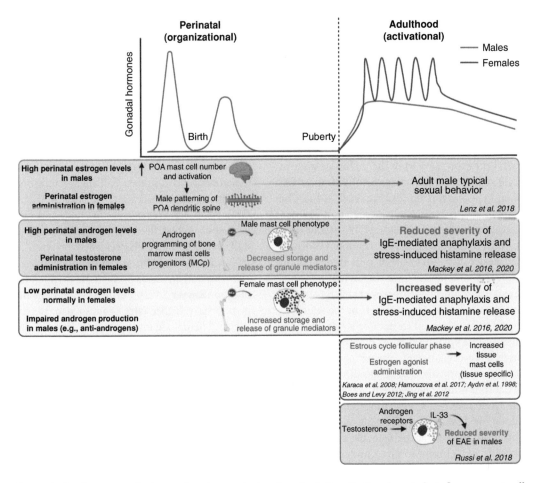

Figure 1. Sex hormones during perinatal organizational and adult activational periods influence mast cell development and activity, shaping neurobehavioral development and mast cell–associated disease across the life span. During the sensitive perinatal period, exposure to sex hormones, either naturally (e.g., male perinatal androgen surge) or potentially a result of endocrine, genetic, and environmental factors, can impact early mast cell activity and subsequent neurobehavioral development, and shape the development and long-term functional response potential of the mast cells that contribute to sex differences in mast cell function and disease risk. Changing sex hormone levels in adulthood (e.g., estrous cycle) alters tissue cell number and activity in various tissues that play a role in normal physiological functions and disease risk. (POA) Preoptic area, (EAE) experimental autoimmune encephalomyelitis.

influence mast cell number and activity, and likely explain some of the sex disparities in mast cell responses, sex biases in mast cell–related disorders are often evident in children prior to puberty, and therefore the study of early life mechanisms driving lifelong changes in development and disease risk is critically important. In addition, understanding how adult sex hormones modulate mast cell function and how perinatal androgens alter the developmental

programming of mast cells and their role in development, host defense, and inflammatory disease risk could offer new insights into host and environmental factors affecting disease risk across the life span and development of novel therapeutic targets to decrease mast cell disease susceptibility. Conversely, given the critical and beneficial role of mast cell activation in pathogen host defense and immunity, deciphering this mechanism may also provide useful targets for enhancing immune responses when that would be beneficial.

ACKNOWLEDGMENTS

This work was supported by National Institutes of Health Grants NIH R01 HD072968-01-05 (to A.J.M.), NIH R01 HD072968-06-10 (to A.J.M.), NIH R21 AI140413 (to A.J.M.), K08 DK084313 (to A.J.M.), NIH F30 OD025354 (to E.M.), and United States Department of Agriculture (USDA) Grant #2019-67015-29483 (to A.J.M.). Figure created with BioRender.com

REFERENCES

Abraham SN, Malaviya R. 1997. Mast cells in infection and immunity. *Infect Immun* **65:** 3501–3508. doi:10.1128/iai .65.9.3501-3508.1997

Acker WW, Plasek JM, Blumenthal KG, Lai KH, Topaz M, Seger DL, Goss FR, Slight SP, Bates DW, Zhou L. 2017. Prevalence of food allergies and intolerances documented in electronic health records. *J Allergy Clin Immunol* **140:** 1587–1591.e1. doi:10.1016/j.jaci.2017.04.006

Alamanos Y, Drosos AA. 2005. Epidemiology of adult rheumatoid arthritis. *Autoimmun Rev* **4:** 130–136. doi:10 .1016/j.autrev.2004.09.002

Anaf V, Chapron C, El Nakadi I, De Moor V, Simonart T, Noel JC. 2006. Pain, mast cells, and nerves in peritoneal, ovarian, and deep infiltrating endometriosis. *Fertil Steril* **86:** 1336–1343. doi:10.1016/j.fertnstert.2006.03.057

Arnold AP, Chen X. 2009. What does the "four core genotypes" mouse model tell us about sex differences in the brain and other tissues? *Front Neuroendocrinol* **30:** 1–9. doi:10.1016/j.yfrne.2008.11.001

Arnold AP, Chen X, Itoh Y. 2012. What a difference an X or Y makes: sex chromosomes, gene dose, and epigenetics in sexual differentiation. *Handb Exp Pharmacol* **214:** 67–88. doi:10.1007/978-3-642-30726-3_4

Aydın Y, Tunçel N, Gürer F, Tunçel M, Koşar M, Oflaz G. 1998. Ovarian, uterine and brain mast cells in female rats: cyclic changes and contribution to tissue histamine. *Comp Biochem Physiol A Mol Integr Physiol* **120:** 255–262. doi:10.1016/S1095-6433(98)00027-0

Ballardini N, Kull I, Söderhäll C, Lilja G, Wickman M, Wahlgren CF. 2013. Eczema severity in preadolescent children and its relation to sex, filaggrin mutations, asthma, rhinitis, aggravating factors and topical treatment: a report from the BAMSE birth cohort. *Br J Dermatol* **168:** 588–594. doi:10.1111/bjd.12196

Berletch JB, Ma W, Yang F, Shendure J, Noble WS, Disteche CM, Deng X. 2015. Escape from X inactivation varies in mouse tissues. *PLoS Genet* **11:** e1005079. doi:10.1371/ journal.pgen.1005079

Berry SH, Elliott MN, Suttorp M, Bogart LM, Stoto MA, Eggers P, Nyberg L, Clemens JQ. 2011. Prevalence of symptoms of bladder pain syndrome/interstitial cystitis among adult females in the United States. *J Urol* **186:** 540–544. doi:10.1016/j.juro.2011.03.132

Boes T, Levy D. 2012. Influence of sex, estrous cycle, and estrogen on intracranial dural mast cells. *Cephalalgia* **32:** 924–931. doi:10.1177/0333102412454947

Bornstein J, Goldschmid N, Sabo E. 2004. Hyperinnervation and mast cell activation may be used as histopathologic diagnostic criteria for vulvar vestibulitis. *Gynecol Obstet Invest* **58:** 171–178. doi:10.1159/000079663

Bradesi S, Eutamene H, Theodorou V, Fioramonti J, Bueno L. 2001. Effect of ovarian hormones on intestinal mast cell reactivity to substance P. *Life Sci* **68:** 1047–1056. doi:10 .1016/S0024-3205(00)01008-0

Branum AM, Lukacs SL. 2008. Food allergy among U.S. children: trends in prevalence and hospitalizations. *NCHS Data Brief* **10:** 1–8.

Brown MA, Hatfield JK. 2012. Mast cells are important modifiers of autoimmune disease: with so much evidence, why is there still controversy? *Front Immunol* **3:** 147. doi:10 .3389/fimmu.2012.00147

Brown MA, Tanzola MB, Robbie-Ryan M. 2002. Mechanisms underlying mast cell influence on EAE disease course. *Mol Immunol* **38:** 1373–1378. doi:10.1016/ S0161-5890(02)00091-3

Carrel L, Willard HF. 2005. X-inactivation profile reveals extensive variability in X-linked gene expression in females. *Nature* **434:** 400–404. doi:10.1038/nature03479

Chen W, Beck I, Schober W, Brockow K, Effner R, Buters JT, Behrendt H, Ring J. 2010. Human mast cells express androgen receptors but treatment with testosterone exerts no influence on IgE-independent mast cell degranulation elicited by neuromuscular blocking agents. *Exp Dermatol* **19:** 302–304. doi:10.1111/j.1600-0625.2009.00969.x

Chiaroni-Clarke RC, Munro JE, Ellis JA. 2016. Sex bias in paediatric autoimmune disease—not just about sex hormones? *J Autoimmun* **69:** 12–23. doi:10.1016/j.jaut.2016 .02.011

Cortez D, Marin R, Toledo-Flores D, Froidevaux L, Liechti A, Waters PD, Grützner F, Kaessmann H. 2014. Origins and functional evolution of Y chromosomes across mammals. *Nature* **508:** 488–493. doi:10.1038/nature13151

D'Costa S, Ayyadurai S, Gibson AJ, Mackey E, Rajput M, Sommerville LJ, Wilson N, Li Y, Kubat E, Kumar A, et al. 2019. Mast cell corticotropin-releasing factor subtype 2 suppresses mast cell degranulation and limits the severity of anaphylaxis and stress-induced intestinal permeability. *J Allergy Clin Immunol* **143:** 1865–1877.e4. doi:10.1016/j .jaci.2018.08.053

Desbiens L, Lapointe C, Gharagozloo M, Mahmoud S, Pejler G, Gris D, D'Orléans-Juste P. 2016. Significant contribution of mouse mast cell protease 4 in early phases of experimental autoimmune encephalomyelitis. *Mediators Inflamm* **2016:** 9797021. doi:10.1155/2016/9797021

Flanagan CE, Wise SK, DelGaudio JM, Patel ZM. 2015. Association of decreased rate of influenza vaccination with increased subjective olfactory dysfunction. *JAMA Otolaryngol Head Neck Surg* **141:** 225–228. doi:10.1001/jamaoto .2014.3399

Foo YZ, Nakagawa S, Rhodes G, Simmons LW. 2017. The effects of sex hormones on immune function: a meta-analysis. *Biol Rev Camb Philos Soc* **92**: 551–571. doi:10 .1111/brv.12243

Fröhlich M, Pinart Gilberga M, Keller T, Reich A, Cabieses B, Hohmann C, Postma DS, Bousquet J, Antó i Boqué JM, Keil T. 2017. Is there a sex-shift in prevalence of allergic rhinitis and comorbid asthma from childhood to adulthood? A meta-analysis. *Clin Transl Allergy* **7**: 44. doi:10 .1186/s13601-017-0176-5

Galli SJ, Tsai M. 2012. IgE and mast cells in allergic disease. *Nat Med* **18**: 693–704. doi:10.1038/nm.2755

Galli SJ, Kalesnikoff J, Grimbaldeston MA, Piliponsky AM, Williams CM, Tsai M. 2005. Mast cells as "tunable" effector and immunoregulatory cells: recent advances. *Annu Rev Immunol* **23**: 749–786. doi:10.1146/annurev .immunol.21.120601.141025

Gebhard C, Regitz-Zagrosek V, Neuhauser HK, Morgan R, Klein SL. 2020. Impact of sex and gender on COVID-19 outcomes in Europe. *Biol Sex Differ* **11**: 1–13. doi:10 .1186/s13293-020-00304-9

Gilfillan AM, Peavy RD, Metcalfe DD. 2009. Amplification mechanisms for the enhancement of antigen-mediated mast cell activation. *Immunol Res* **43**: 15–24. doi:10 .1007/s12026-008-8046-9

Hamouzova P, Cizek P, Novotny R, Bartoskova A, Tichy F. 2017. Distribution of mast cells in the feline ovary in various phases of the oestrous cycle. *Reprod Domest Anim* **52**: 483–486. doi:10.1111/rda.12938

Hamouzova P, Cizek P, Bartoskova A, Vitasek R, Tichy F. 2020. Changes in the mast cell distribution in the canine ovary and uterus throughout the oestrous cycle. *Reprod Domest Anim* **55**: 479–485. doi:10.1111/rda.13641

Hiort O. 2013. The differential role of androgens in early human sex development. *BMC Med* **11**: 152. doi:10.1186/ 1741-7015-11-152

Hofmann AM, Abraham SN. 2009. New roles for mast cells in modulating allergic reactions and immunity against pathogens. *Curr Opin Immunol* **21**: 679–686. doi:10 .1016/j.coi.2009.09.007

Hohmann C, Keller T, Gehring U, Wijga A, Standl M, Kull I, Bergstrom A, Lehmann I, von Berg A, Heinrich J, et al. 2019. Sex-specific incidence of asthma, rhinitis and respiratory multimorbidity before and after puberty onset: individual participant meta-analysis of five birth cohorts collaborating in MeDALL. *BMJ Open Respir Res* **6**: e000460. doi:10.1136/bmjresp-2019-000460

Hox V, Desai A, Bandara G, Gilfillan AM, Metcalfe DD, Olivera A. 2015. Estrogen increases the severity of anaphylaxis in female mice through enhanced endothelial nitric oxide synthase expression and nitric oxide production. *J Allergy Clin Immunol* **135**: 729–736.e5. doi:10.1016/j.jaci.2014.11.003

Ishii T, Fujita T, Matsushita T, Yanaba K, Hasegawa M, Nakashima H, Ogawa F, Shimizu K, Takehara K, Tedder TF, et al. 2009. Establishment of experimental eosinophilic vasculitis by IgE-mediated cutaneous reverse passive Arthus reaction. *Am J Pathol* **174**: 2225–2233. doi:10 .2353/ajpath.2009.080223

Jensen F, Woudwyk M, Teles A, Woidacki K, Taran F, Costa S, Malfertheiner SF, Zenclussen AC. 2010. Estradiol and progesterone regulate the migration of mast cells from the periphery to the uterus and induce their maturation and

degranulation. *PLoS ONE* **5**: e14409. doi:10.1371/journal .pone.0014409

Jing H, Wang Z, Chen Y. 2012. Effect of oestradiol on mast cell number and histamine level in the mammary glands of rat. *Anat Histol Embryol* **41**: 170–176. doi:10.1111/j .1439-0264.2011.01120.x

Kaczmarczyk-Sekuła K, Dyduch G, Kostański M, Wielowieyska-Szybińska D, Szpor J, Białas M, Okoń K. 2015. Mast cells in systemic and cutaneous lupus erythematosus. *Pol J Pathol* **66**: 397–402. doi:10.5114/pjp.2015.57253

Karaca T, Arikan S, Kalender H, Yoruk M. 2008. Distribution and heterogeneity of mast cells in female reproductive tract and ovary on different days of the oestrus cycle in Angora goats. *Reprod Domest Anim* **43**: 451–456. doi:10 .1111/j.1439-0531.2007.00934.x

Kashikar-Zuck S, Ting TV. 2014. Juvenile fibromyalgia: current status of research and future developments. *Nat Rev Rheumatol* **10**: 89–96. doi:10.1038/nrrheum.2013.177

Kelly C, Gangur V. 2009. Sex disparity in food allergy: evidence from the PubMed database. *J Allergy (Cairo)* **2009**: 159845. doi:10.1155/2009/159845

Kim MS, Chae HJ, Shin TY, Kim HM, Kim HR. 2001. Estrogen regulates cytokine release in human mast cells. *Immunopharmacol Immunotoxicol* **23**: 495–504. doi:10 .1081/IPH-100108596

King S, Chambers CT, Huguet A, MacNevin RC, McGrath PJ, Parker L, MacDonald AJ. 2011. The epidemiology of chronic pain in children and adolescents revisited: a systematic review. *Pain* **152**: 2729–2738. doi:10.1016/j.pain .2011.07.016

Klein SL. 2004. Hormonal and immunological mechanisms mediating sex differences in parasite infection. *Parasite Immunol* **26**: 247–264. doi:10.1111/j.0141-9838.2004 .00710.x

Klein SL, Flanagan KL. 2016. Sex differences in immune responses. *Nat Rev Immunol* **16**: 626–638. doi:10.1038/ nri.2016.90

Konttinen YT, Hietanen J, Virtanen I, Ma J, Sorsa T, Xu JW, Williams NP, Manthorpe R, Janin A. 2000. Mast cell derangement in salivary glands in patients with Sjögren's syndrome. *Rheumatol Int* **19**: 141–147. doi:10.1007/ s002960050118

Kool B, Chandra D, Fitzharris P. 2016. Adult food-induced anaphylaxis hospital presentations in New Zealand. *Postgrad Med J* **92**: 640–644. doi:10.1136/postgradmedj-2015-133530

Korterink JJ, Diederen K, Benninga MA, Tabbers MM. 2015. Epidemiology of pediatric functional abdominal pain disorders: a meta-analysis. *PLoS ONE* **10**: e0126982. doi:10 .1371/journal.pone.0126982

Krüger PG, Bø L, Myhr KM, Karlsen AE, Taule A, Nyland HI, Mørk S. 1990. Mast cells and multiple sclerosis: a light and electron microscopic study of mast cells in multiple sclerosis emphasizing staining procedures. *Acta Neurol Scand* **81**: 31–36. doi:10.1111/j.1600-0404.1990.tb0 0927.x

Krystel-Whittemore M, Dileepan KN, Wood JG. 2015. Mast cell: a multi-functional master cell. *Front Immunol* **6**: 620. doi:10.3389/fimmu.2015.00620

Laffont S, Blanquart E, Savignac M, Cénac C, Laverny G, Metzger D, Girard JP, Belz GT, Pelletier L, Seillet C, et al. 2017. Androgen signaling negatively controls group 2 innate lymphoid cells. *J Exp Med* **214:** 1581–1592. doi:10 .1084/jem.20161807

Lenz KM, Pickett LA, Wright CL, Davis KT, Joshi A, McCarthy MM. 2018. Mast cells in the developing brain determine adult sexual behavior. *J Neurosci* **38:** 8044– 8059. doi:10.1523/JNEUROSCI.1176-18.2018

Lenz KM, Pickett LA, Wright CL, Galan A, McCarthy MM. 2019. Prenatal allergen exposure perturbs sexual differentiation and programs lifelong changes in adult social and sexual behavior. *Sci Rep* **9:** 4837. doi:10.1038/s41598-019-41258-2

Libert C, Dejager L, Pinheiro I. 2010. The X chromosome in immune functions: when a chromosome makes the difference. *Nat Rev Immunol* **10:** 594–604. doi:10.1038/nri2815

Ligeiro de Oliveira AP, Oliveira-Filho RM, da Silva ZL, Borelli P, Tavares de Lima W. 2004. Regulation of allergic lung inflammation in rats: interaction between estradiol and corticosterone. *Neuroimmunomodulation.* **11:** 20– 27. doi:10.1159/000072965

Mackey E, Ayyadurai S, Pohl CS, D'Costa S, Li Y, Moeser AJ. 2016. Sexual dimorphism in the mast cell transcriptome and the pathophysiological responses to immunological and psychological stress. *Biol Sex Differ* **7:** 60. doi:10 .1186/s13293-016-0113-7

Mackey E, Thelen KM, Bali V, Fardisi M, Trowbridge M, Jordan CL, Moeser AJ. 2020. Perinatal androgens organize sex differences in mast cells and attenuate anaphylaxis severity into adulthood. *Proc Natl Acad Sci* **117:** 23751–23761. doi:10.1073/pnas.1915075117

Malaviya R, Ikeda T, Ross E, Abraham SN. 1996. Mast cell modulation of neutrophil influx and bacterial clearance at sites of infection through TNF-α. *Nature* **381:** 77–80. doi:10.1038/381077a0

McDermott JR, Bartram RE, Knight PA, Miller HR, Garrod DR, Grencis RK. 2003. Mast cells disrupt epithelial barrier function during enteric nematode infection. *Proc Natl Acad Sci* **100:** 7761–7766. doi:10.1073/pnas.1231488100

McGowan EC, Keet CA. 2013. Prevalence of self-reported food allergy in the National Health and Nutrition Examination Survey (NHANES) 2007–2010. *J Allergy Clin Immunol* **132:** 1216–1219.e5. doi:10.1016/j.jaci.2013.07.018

McKean PG, Pritchard DI. 1989. The action of a mast cell protease on the cuticular collagens of *Necator americanus. Parasite Immunol* **11:** 293–297. doi:10.1111/j .1365-3024.1989.tb00667.x

Meleine M, Matricon J. 2014. Gender-related differences in irritable bowel syndrome: potential mechanisms of sex hormones. *World J Gastroenterol* **20:** 6725–6743. doi:10 .3748/wjg.v20.i22.6725

Mogil JS. 2012. Sex differences in pain and pain inhibition: multiple explanations of a controversial phenomenon. *Nat Rev Neurosci* **13:** 859–866. doi:10.1038/nrn3360

Moloney RD, Sajjad J, Foley T, Felice VD, Dinan TG, Cryan JF, O'Mahony SM. 2016. Estrous cycle influences excitatory amino acid transport and visceral pain sensitivity in the rat: effects of early-life stress. *Biol Sex Differ* **7:** 33. doi:10.1186/s13293-016-0086-6

Muñoz-Cruz S, Mendoza-Rodríguez Y, Nava-Castro KE, Yepez-Mulia L, Morales-Montor J. 2015. Gender-related effects of sex steroids on histamine release and FcεRI

expression in rat peritoneal mast cells. *J Immunol Res* **2015:** 351829. doi:10.1155/2015/351829

Narita S, Goldblum RM, Watson CS, Brooks EG, Estes DM, Curran EM, Midoro-Horiuti T. 2007. Environmental estrogens induce mast cell degranulation and enhance IgE-mediated release of allergic mediators. *Environ Health Perspect* **115:** 48–52. doi:10.1289/ehp.9378

Nigrovic PA, Binstadt BA, Monach PA, Johnsen A, Gurish M, Iwakura Y, Benoist C, Mathis D, Lee DM. 2007. Mast cells contribute to initiation of autoantibody-mediated arthritis via IL-1. *Proc Natl Acad Sci* **104:** 2325–2330. doi:10.1073/pnas.0610852103

Orinska Z, Bulanova E, Budagian V, Metz M, Maurer M, Bulfone-Paus S. 2005. TLR3-induced activation of mast cells modulates CD8⁺ T-cell recruitment. *Blood* **106:** 978– 987. doi:10.1182/blood-2004-07-2656

Orton SM, Herrera BM, Yee IM, Valdar W, Ramagopalan SV, Sadovnick AD, Ebers GC; Canadian Collaborative Study Group. 2006. Sex ratio of multiple sclerosis in Canada: a longitudinal study. *Lancet Neurol* **5:** 932–936. doi:10.1016/S1474-4422(06)70581-6

Osman M, Hansell AL, Simpson CR, Hollowell J, Helms PJ. 2007a. Gender-specific presentations for asthma, allergic rhinitis and eczema in primary care. *Prim Care Respir J* **16:** 28–35. doi:10.3132/pcrj.2007.00006

Osman M, Tagiyeva N, Wassall HJ, Ninan TK, Devenny AM, McNeill G, Helms PJ, Russell G. 2007b. Changing trends in sex specific prevalence rates for childhood asthma, eczema, and hay fever. *Pediatr Pulmonol* **42:** 60–65. doi:10.1002/ppul.20545

Özen A, Ergün L, Ergün E, ŞimŞek N. 2007. Morphological studies on ovarian mast cells in the cow. *Turk J Vet Anim Sci* **31:** 131–136.

Pang X, Cotreau-Bibbo MM, Sant GR, Theoharides TC. 1995. Bladder mast cell expression of high affinity oestrogen receptors in patients with interstitial cystitis. *Br J Urol* **75:** 154–161. doi:10.1111/j.1464-410X.1995.tb07303.x

Pawankar R, Canonica GW, Holgate ST, Lockey RF. 2012. Allergic diseases and asthma: a major global health concern. *Curr Opin Allergy Clin Immunol* **12:** 39–41. doi:10 .1097/ACI.0b013e32834ec13b

Pinart M, Keller T, Reich A, Fröhlich M, Cabieses B, Hohmann C, Postma DS, Bousquet J, Antó JM, Keil T. 2017. Sex-related allergic rhinitis prevalence switch from childhood to adulthood: a systematic review and meta-analysis. *Int Arch Allergy Immunol* **172:** 224–235. doi:10.1159/000464324

Poulos LM, Waters AM, Correll PK, Loblay RH, Marks GB. 2007. Trends in hospitalizations for anaphylaxis, angioedema, and urticaria in Australia, 1993–1994 to 2004– 2005. *J Allergy Clin Immunol* **120:** 878–884. doi:10 .1016/j.jaci.2007.07.040

Rozniecki JJ, Hauser SL, Stein M, Lincoln R, Theoharides TC. 1995. Elevated mast cell tryptase in cerebrospinal fluid of multiple sclerosis patients. *Ann Neurol* **37:** 63– 66. doi:10.1002/ana.410370112

Russi AE, Ebel ME, Yang Y, Brown MA. 2018. Male-specific IL-33 expression regulates sex-dimorphic EAE susceptibility. *Proc Natl Acad Sci* **115:** E1520–E1529. doi:10.1073/pnas.1710401115

Saini SS, Paterniti M, Vasagar K, Gibbons SP Jr, Sterba PM, Vonakis BM. 2009. Cultured peripheral blood mast cells from chronic idiopathic urticaria patients spontaneously degranulate upon IgE sensitization: relationship to expression of Syk and SHIP-2. *Clin Immunol* **132:** 342–348. doi:10.1016/j.clim.2009.05.003

Sawyer CC. 2012. Child mortality estimation: estimating sex differences in childhood mortality since the 1970s. *PLoS Med* **9:** e1001287. doi:10.1371/journal.pmed.1001287

Sayed BA, Christy AL, Walker ME, Brown MA. 2010. Meningeal mast cells affect early T cell central nervous system infiltration and blood–brain barrier integrity through TNF: a role for neutrophil recruitment? *J Immunol* **184:** 6891–6900. doi:10.4049/jimmunol.1000126

Shelburne CP, Nakano H, St. John AL, Chan C, McLachlan JB, Gunn MD, Staats HF, Abraham SN. 2009. Mast cells augment adaptive immunity by orchestrating dendritic cell trafficking through infected tissues. *Cell Host Microbe* **6:** 331–342. doi:10.1016/j.chom.2009.09.004

Siebenhaar F, Syska W, Weller K, Magerl M, Zuberbier T, Metz M, Maurer M. 2007. Control of *Pseudomonas aeruginosa* skin infections in mice is mast cell-dependent. *Am J Pathol* **170:** 1910–1916. doi:10.2353/ajpath.2007.060770

Sperber AD, Dumitrascu D, Fukudo S, Gerson C, Ghoshal UC, Gwee KA, Hungin APS, Kang JY, Minhu C, Schmulson M, et al. 2017. The global prevalence of IBS in adults remains elusive due to the heterogeneity of studies: a Rome Foundation working team literature review. *Gut* **66:** 1075–1082. doi:10.1136/gutjnl-2015-311240

Stovner L, Hagen K, Jensen R, Katsarava Z, Lipton R, Scher A, Steiner T, Zwart JA. 2007. The global burden of headache: a documentation of headache prevalence and disability worldwide. *Cephalalgia* **27:** 193–210. doi:10.1111/j.1468-2982.2007.01288.x

Sullivan KE. 2007. Inflammation in juvenile idiopathic arthritis. *Rheum Dis Clin North Am* **33:** 365–388. doi:10.1016/j.rdc.2007.07.004

Sutherland RE, Olsen JS, McKinstry A, Villalta SA, Wolters PJ. 2008. Mast cell IL-6 improves survival from *Klebsiella* pneumonia and sepsis by enhancing neutrophil killing. *J Immunol* **181:** 5598–5605. doi:10.4049/jimmunol.181.8.5598

Tiuria R, Horii Y, Tateyama S, Tsuchiya K, Nawa Y. 1994. The Indian soft-furred rat, *Millardia meltada*, a new host for *Nippostrongylus brasiliensis*, showing androgen-dependent sex difference in intestinal mucosal defence. *Int J Parasitol* **24:** 1055–1057. doi:10.1016/0020-7519(94)90170-8

Tiuria R, Horii Y, Makimura S, Ishikawa N, Tsuchiya K, Nawa Y. 1995. Effect of testosterone on the mucosal defence against intestinal helminths in Indian soft-furred rats, *Millardia meltada* with reference to goblet and mast cell responses. *Parasite Immunol* **17:** 479–484. doi:10.1111/j.1365-3024.1995.tb00918.x

Vasiadi M, Kempuraj D, Boucher W, Kalogeromitros D, Theoharides TC. 2006. Progesterone inhibits mast cell secretion. *Int J Immunopathol Pharmacol* **19:** 787–794. doi:10.1177/039463200601900408

Vaz GT, Vasconcelos MM, Oliveira EA, Ferreira AL, Magalhães PG, Silva FM, Lima EM. 2012. Prevalence of lower urinary tract symptoms in school-age children. *Pediatr Nephrol* **27:** 597–603. doi:10.1007/s00467-011-2028-1

Vermillion DL, Ernst PB, Scicchitano R, Collins SM. 1988. Antigen-induced contraction of jejunal smooth muscle in the sensitized rat. *Am J Physiol* **255:** G701–G708. doi:10.1152/ajpgi.1988.255.6.G701

Victor TW, Hu X, Campbell JC, Buse DC, Lipton RB. 2010. Migraine prevalence by age and sex in the United States: a life-span study. *Cephalalgia* **30:** 1065–1072. doi:10.1177/0333102409355601

Voehringer D. 2013. Protective and pathological roles of mast cells and basophils. *Nat Rev Immunol* **13:** 362–375. doi:10.1038/nri3427

vom Steeg LG, Klein SL. 2016. SeXX matters in infectious disease pathogenesis. *PLoS Pathog* **12:** e1005374. doi:10.1371/journal.ppat.1005374

Vrieze A, Postma DS, Kerstjens HA. 2003. Perimenstrual asthma: a syndrome without known cause or cure. *J Allergy Clin Immunol* **112:** 271–282. doi:10.1067/mai.2003.1676

Walitt B, Nahin RL, Katz RS, Bergman MJ, Wolfe F. 2015. The prevalence and characteristics of fibromyalgia in the 2012 National Health Interview Survey. *PLoS ONE* **10:** e0138024. doi:10.1371/journal.pone.0138024

Webb LM, Lieberman P. 2006. Anaphylaxis: a review of 601 cases. *Ann Allergy Asthma Immunol* **97:** 39–43. doi:10.1016/S1081-1206(10)61367-1

Wei OL, Hilliard A, Kalman D, Sherman M. 2005. Mast cells limit systemic bacterial dissemination but not colitis in response to *Citrobacter rodentium*. *Infect Immun* **73:** 1978–1985. doi:10.1128/IAI.73.4.1978-1985.2005

Wijchers PJ, Festenstein RJ. 2011. Epigenetic regulation of autosomal gene expression by sex chromosomes. *Trends Genet* **27:** 132–140. doi:10.1016/j.tig.2011.01.004

Wintroub BU, Mihm MC Jr, Goetzl EJ, Soter NA, Austen KF. 1978. Morphologic and functional evidence for release of mast-cell products in bullous pemphigoid. *N Engl J Med* **298:** 417–421. doi:10.1056/NEJM197802232980803

Wong GW, Zhuo L, Kimata K, Lam BK, Satoh N, Stevens RL. 2014. Ancient origin of mast cells. *Biochem Biophys Res Commun* **451:** 314–318. doi:10.1016/j.bbrc.2014.07.124

Xu S, Wang X, Zhao J, Yang S, Dong L, Qin B. 2020. GPER-mediated, oestrogen-dependent visceral hypersensitivity in stressed rats is associated with mast cell tryptase and histamine expression. *Fundam Clin Pharmacol* **34:** 433–443. doi:10.1111/fcp.12537

Zaitsu M, Narita S, Lambert KC, Grady JJ, Estes DM, Curran EM, Brooks EG, Watson CS, Goldblum RM, Midoro-Horiuti T. 2007. Estradiol activates mast cells via a non-genomic estrogen receptor-α and calcium influx. *Mol Immunol* **44:** 1977–1985. doi:10.1016/j.molimm.2006.09.030

Zhu TH, Ding SJ, Li TT, Zhu LB, Huang XF, Zhang XM. 2018. Estrogen is an important mediator of mast cell activation in ovarian endometriomas. *Reproduction* **155:** 73–83. doi:10.1530/REP-17-0457

Cite this article as *Cold Spring Harb Perspect Biol* doi: 10.1101/cshperspect.a039172

Sex Differences in Spotted Hyenas

S. Kevin McCormick,[1,2] Kay E. Holekamp,[1,2] Laura Smale,[1,2,3] Mary L. Weldele,[4] Stephen E. Glickman,[4,6] and Ned J. Place[5]

[1]Department of Integrative Biology, Michigan State University, East Lansing, Michigan 48824, USA

[2]Ecology, Evolution, and Behavior Program, Michigan State University, East Lansing, Michigan 48824, USA

[3]Department of Psychology, Michigan State University, East Lansing, Michigan 48824, USA

[4]Departments of Psychology and Integrative Biology, University of California, Berkeley, California 94720, USA

[5]Department of Population Medicine and Diagnostic Sciences, College of Veterinary Medicine, Cornell University, Ithaca, New York 14853, USA

Correspondence: holekamp@msu.edu

The apparent virilization of the female spotted hyena raises questions about sex differences in behavior and morphology. We review these sex differences to find a mosaic of dimorphic traits, some of which conform to mammalian norms. These include space-use, dispersal behavior, sexual behavior, and parental behavior. By contrast, sex differences are reversed from mammalian norms in the hyena's aggressive behavior, social dominance, and territory defense. Androgen exposure early in development appears to enhance aggressiveness in female hyenas. Weapons, hunting behavior, and neonatal body mass do not differ between males and females, but females are slightly larger than males as adults. Sex differences in the hyena's nervous system are relatively subtle. Overall, it appears that the "masculinized" behavioral traits in female spotted hyenas are those, such as aggression, that are essential to ensuring consistent access to food; food critically limits female reproductive success in this species because female spotted hyenas have the highest energetic investment per litter of any mammalian carnivore. Evidently, natural selection has acted to modify traits related to food access, but has left intact those traits that are unrelated to acquiring food, such that they conform to patterns of sexual dimorphism in other mammals.

Spotted hyenas (*Crocuta crocuta*) are large mammalian carnivores that occur throughout sub-Saharan Africa. They show many unique and fascinating characteristics, such as living in large, complex societies in which they must compete and cooperate with non-kin as well as kin. However, the aspect of the biology of spotted hyenas that many people find most intriguing is the apparent "masculinization" of females. This female virilization manifests itself in several obvious ways. For example, in contrast to the situation characteristic of most other mammalian species, female spotted hyenas are larger than males, they are socially dominant to

---

[6]Deceased.

males, and they show genitalia that are astonishingly male-like. These "sex-role-reversed" traits coexist in the same individuals with other sexually dimorphic traits that are much like those shown by virtually all "typical" female mammals, such as dogs, antelope, rats, and baboons. The hyena's chimeric blend of feminine and "masculinized" traits is particularly intriguing because it raises so many questions about how and why some traits have been "masculinized" in this species, whereas others have not. Here, we summarize existing information about sexually dimorphic traits in the behavior and morphology of spotted hyenas. We will consider sexually dimorphic traits in adults of this species, including both those that are "reversed" from patterns found in most other mammals and those known to show the same patterns of dimorphism typical of other mammals. Wherever possible, we also briefly review what is known about the development of sexually dimorphic traits in spotted hyenas. Finally, we discuss the adaptive significance of sexually dimorphic traits in this species.

SEX DIFFERENCES IN BEHAVIOR

Sex Differences in Space Use, Territory Defense, and Dispersal

Spotted hyenas live in stable social groups, called "clans," which may contain up to 130 individuals. They defend group territories against encroachment by neighboring conspecifics. They engage in border patrols and "wars" with neighboring clans. In contrast to many other mammals that defend group territories, female hyenas initiate and lead most territorial advertisement and defense efforts. Females are more likely than males to lead border patrols and clan wars, and females tend to scent-mark along territorial boundaries at higher hourly rates than do adult males (Henschel and Skinner 1991; Boydston et al. 2001). Thus, female hyenas are willing to assume more risks and expend more energy during territorial defense than are males. These sex differences are consistent with the hypothesis that male and female clan members derive different selective benefits from advertisement and defense of group territories. Defense of food re-

sources appears to be the primary function of territoriality in spotted hyenas (Kruuk 1972; Henschel and Skinner 1991). Indeed, it appears that natural selection has favored female spotted hyenas to maintain boundaries of a territory that supports enough herbivore prey to feed themselves and their young throughout the year.

In other respects, the territorial and space-use behavior of spotted hyenas resembles that of other territorial mammals. For instance, during territorial encounters between residents and aliens, residents are more likely to attack same-sex than opposite sex intruders (Boydston et al. 2001). As also occurs in most other social mammals, male spotted hyenas disperse from their natal clans after they reach reproductive maturity, whereas females are philopatric and spend their entire lives in the natal clan (Smale et al. 1997; Höner et al. 2007). No measures of space use are sexually dimorphic among young hyenas until these animals are ~30 mo of age, which is roughly 6 mo after they reach reproductive maturity. Late in the third year of life, males start making exploratory excursions into the territories of neighboring clans, whereas females do not; these sex differences in space-use persist throughout the remainder of the life span. Males are found farther from the geographic center of their natal territory than are females, and the mean size of individual home ranges is larger for males than females (Boydston et al. 2005). As adults, male spotted hyenas travel 17.7 km/d, whereas females travel only 12.36 km/d (Kolowski et al. 2007).

Sex Differences in Reproductive Behavior

Although copulation is rather challenging for male spotted hyenas because of the female's unusual external genitalia (see below), otherwise spotted hyenas show the same suite of sex differences in reproductive behavior as those typical of other mammals. For instance, males approach females at higher rates than vice versa (Szykman et al. 2007), although, in contrast to most other male mammals, male spotted hyenas appear to be extremely nervous when courting females. Copulation involves the male mounting the female and inserting his erect penis into the female's flaccid clitoris. Both males and females mate promiscuously (Szykman et al. 2007).

Hyenas are born in litters that usually contain only one or two cubs, but maternal investment in each cub is enormous. As occurs in most mammals, female spotted hyenas do all the parenting; the low social status of sires effectively prevents them from being able to assist their offspring even to the same small extent as males can assist their young in other polygynous, group-living mammals (e.g., baboons; Buchan et al. 2003). Spotted hyena cubs rely exclusively on their mother's milk during the first 6 mo of life, and although they then start eating some solid food, cubs continue to rely largely on milk until they are weaned, which typically occurs at 12–20 mo of age (Kruuk 1972; Hofer and East 1995; Holekamp et al. 1997); this represents an extremely protracted lactation period relative to those of other carnivores of similar body mass. For example, canid and felid species the same size as, or larger than, spotted hyenas wean their young when they are only 1.1 to 6 mo of age, including species as large as wolves, lions, and tigers (Watts et al. 2009). The milk produced by spotted hyenas is also unusually rich; it has the highest protein content of milk from any fissiped carnivore (Hofer and East 1996), a fat content exceeded only by that of milk produced by palearctic bears and sea otters, and a higher gross energy density than the milk of most other terrestrial carnivores (Hofer and East 1995). Because of the high energy content of their milk and the long period of lactation, spotted hyenas have the highest energetic investment per litter of any mammalian carnivore (Oftedal and Gittleman 1989; Hofer et al. 2016). It follows that two critical factors affecting cub growth and survival are maternal access to food (Swanson et al. 2011; Holekamp and Strauss 2020) and nursing frequency (Hofer et al. 2016), both of which vary with maternal social rank.

Sex Differences in Aggression and Dominance among Adults

Adult female spotted hyenas are socially dominant to all adult males not born in the females' natal clan, so females can straightforwardly dis-

place immigrant males from desired resources such as food (Kruuk 1972; Mills 1990). Natal animals of both sexes acquire ranks immediately below those of their mothers via a prolonged learning process early in development, so they can dominate all immigrant males. However, most males disperse before breeding and behave submissively to all new hyenas encountered outside the natal territory. Thus, the mechanisms by which adult social rank is acquired differ between male and female hyenas. Females maintain their natal ranks as long as they live in the natal clan; females do this largely by behaving aggressively to lower-ranking clan mates. In contrast, when males disperse to new clans, they assume the lowest possible rank in the new clan, where they follow a queuing convention in which the most recent immigrant is the lowest-ranking animal in the entire clan (Smale et al. 1997; East and Hofer 2001). Males only improve their status when higher-ranking immigrants die or engage in secondary dispersal, which occurs in roughly 40% of males (Van Horn et al. 2003).

Although some have argued that female dominance among spotted hyenas is strictly the result of more social support for females than males (e.g., Vullioud et al. 2019), compelling evidence indicates that behaviors associated with the acquisition and maintenance of social rank are strongly sexually dimorphic in this species, and in fact require no social support. Female spotted hyenas emit aggressive acts at higher rates than do males (Fig. 1A), they emit more intense aggressive acts (Fig. 1B), they are more tenacious fighters (Fig. 2), and they show unambiguous "role-reversed" sex differences in aggressive behavior from a very young age. Thus, in contrast to most mammals, female spotted hyenas are substantially more aggressive than males (Szykman et al. 2003; McCormick and Holekamp, in review). This notion is also supported by wounding data obtained from the several hundred spotted hyenas we have immobilized in Kenya (Fig. 3). These data show clearly that adult females bear many more wounds, on average, than do adult males; this is the opposite of the pattern found in most other mammals (e.g., primates [Holekamp 1984]; rodents

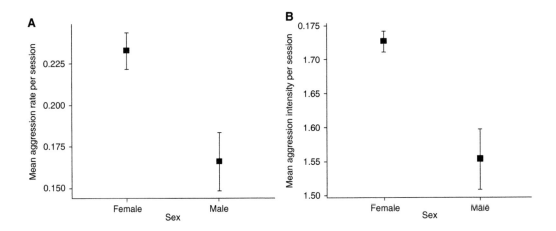

Figure 1. Sex differences in the mean (*A*) rates of emission of aggressive acts by adult hyenas per observation session, and (*B*) intensity of aggressive behaviors emitted by adult natal females and immigrant males, controlled for social rank, time observed, and immigration status. Sampled hyenas include 57 adult immigrant males and 128 adult females. Error bars represent 95% confidence intervals.

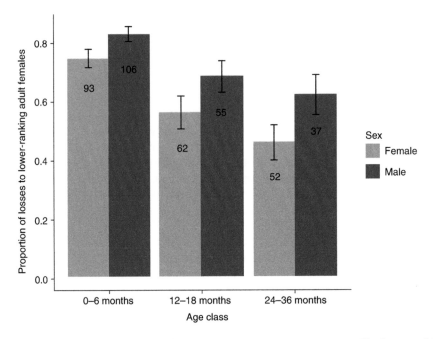

Figure 2. Sex differences in losses by juvenile hyenas in aggressive interactions initiated by lower-ranking adult females. Bars represent the proportion of fights lost per individual in each sex and age class, error bars represent standard error of the mean, and the numbers above the error bars represent the number of individuals included. Females are represented by pale gray bars, and males by dark gray bars. Significant differences were found based on sex ($F = 9.71$, $P = 0.002$) and age class ($F = 19.05$, $P < 0.001$). Modified from Smale et al. (1993).

     Cite this article as *Cold Spring Harb Perspect Biol* doi: 10.1101/cshperspect.a039180

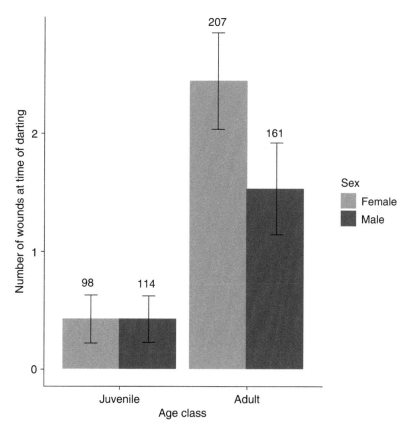

Figure 3. Sex differences in wounding among juvenile (pale gray, younger than 24 mo) and adult (dark gray, older than 24 mo) spotted hyenas of each sex. Numbers above bars indicate individuals sampled. Significant differences were found based on sex ($F = 5.17$, $P = 0.024$), age class ($F = 72.34$, $P < 0.001$), and an interaction between sex and age class ($F = 6.64$, $P = 0.01$).

[Pusey and Packer 1978]). Interestingly, the sex difference in wounding among hyenas does not emerge until adulthood, as males and females 24 mo of age or less do not differ in wounding frequency (Fig. 3). If most wounds were caused by prey animals during capture, we would expect male and female hyenas to show similar rates of wounding because adults of both sexes are equally successful at hunting their own prey (Holekamp et al. 1997). However, the finding that females sustain so many more wounds than males is consistent with the idea that most wounding occurs during fights over carcasses. Males almost invariably defer to females at kills, whereas other females often do not (Kruuk 1972), and fights among males are typically far less intense than those among females. Our

wounding data are consistent with both of these behavioral tendencies.

Ontogenetic Development of Sex Differences in Aggression

Sex differences in aggressive behavior emerge during the first days of life in young spotted hyenas. Newborn spotted hyenas often fight vigorously with their siblings during the first days or weeks after birth to establish intralitter ranks (Frank et al. 1991; Smale et al. 1999; Wachter et al. 2002; Wahaj and Holekamp 2006). Once intralitter ranks have been established, the rates and intensities at which siblings fight decline; intralitter aggression rates also decline with increasing maternal rank (Golla et al. 1999; Smale et al. 1999). When litters are of mixed sex, females dominate their male siblings 67% to 84% of the time (Smale et al. 1995; Golla et al. 1999; Wahaj and Holekamp 2006; Benhaiem et al. 2012). Thus, female spotted hyenas evidently come into the world behaving more aggressively than their male peers.

This trend persists throughout ontogeny. The proportion of dyadic fights with lower-born adult females won by hyenas 6–36 mo of age is significantly greater for young females than for their male peers (Smale et al. 1993). Subadult females also dominate lower-born adult females more consistently than do subadult males. That is, although young males can often displace lower-born adult females from desired resources, their rank relationships with lower-born adult females often remain unstable until males disperse from their natal clans (Smale et al. 1993). Furthermore, juvenile females are more persistent than their male peers in their attempts to outrank adult females. For instance, juvenile females are more likely than their male peers to counterattack lower-born females who attack them (Fig. 2; also see Smale et al. 1993). Thus, although aggression rates come to be strongly affected by social rank in adulthood (McCormick and Holekamp, in review), sex differences in aggressive behavior are apparent from birth in this species.

Exposure to androgens early in development appears to enhance aggressiveness in female spotted hyenas. In the wild, both juvenile and adult females whose mothers have higher androgen concentrations during gestation are considerably more aggressive than are same-age females exposed to lower androgen concentrations in utero (Dloniak et al. 2006; Holekamp et al. 2013). Furthermore, experimental exposure to anti-androgens (AAs) during development in utero reduced female aggressiveness later in life among captive hyenas at the Berkeley hyena colony. Pregnant females were treated throughout gestation with a cocktail of AAs (flutamide and finasteride), and the behavior of the offspring from those pregnancies was assessed throughout development. For many years after the AA-treated hyenas reached adulthood, investigators were unable to identify any obvious effects of the prenatal AA treatment on aggressive behavior. In particular, AA-treated females were always the winners over control and AA-treated males in single-bone dyadic tests, wherein two hyenas compete for a single bone tethered to a fence within an enclosure, as described by Beach et al. (1982). However, as the Berkeley Hyena Project neared its end, the bone dyad test was modified to include two bones. Only two double-bone dyad tests were completed before the colony closed, one with a control female and the other with an AA-treated female, in which each was paired with an untreated male. In their single-bone dyad tests with these same males, both females always won, securing all bones for themselves by aggressively displacing the males from proximity to the tethered bones. However, whereas the control female secured both bones for herself in the double-bone test while keeping the male at bay, the AA-treated female allowed the male to have the second bone and feed on it beside her. No untreated captive females or females in the wild ever share food with unrelated males, so even with a sample size of only one, the experimental treatment effects here were highly suggestive. A video showing both control and treatment trials is available (see online Movie S1).

The striking sex differences seen in aggression and dominance in spotted hyenas have clear adaptive value, as higher rates of aggressive behavior enhance reproductive success among females (Watts et al. 2009; Yoshida et al. 2016; McCormick and Holekamp, in review) but not among males (East and Hofer 2001). Furthermore, the establishment of dominance over other females has a much greater influence on the fitness of females than males. Similarly, although rank reversals among adult females are rare in this species, improving one's social status via rank reversals is far more critical to females than males, as the effects of rank reversals are amplified in later generations (Strauss and Holekamp 2019). The uniquely heavy energetic demands of lactation in this species (Oftedal and Gittleman 1989) cause improved access to food resources, often accomplished via aggressive displacement of group-mates from carcasses, to be far more important for female than males (Holekamp and Strauss 2020), and females are clearly willing to fight to maintain or improve their priority of access to food.

SEX DIFFERENCES IN MORPHOLOGY

Body Size and Shape

Sexual size dimorphism is common among mammals; in most species, including most other mammalian carnivores, males on average are larger than females. In contrast, the spotted hyena is one of the rare species in which females are generally larger than males. We took 14 different body measurements from several hundred wild hyenas in Kenya, and found that, although many body size measures differ only by 1% to 5% between the sexes, and although distributions of most size measures overlap for males and females, these sex differences are strongly statistically significant (Swanson et al. 2013). The largest sex differences in body size in spotted hyenas appear in measures of head and neck circumferences, body mass and girth, indicating that adult females are roughly 10% brawnier than males. These traits are larger in adult female hyenas than adult males even when the two sexes are fed identical diets while housed alone throughout development in captivity, allowing us to rule out a strictly environmental explanation for this dimorphism (Swanson et al. 2013). Because the fundamental frequencies of some hyena vocalizations vary with girth measurements and because girth is strongly sexually dimorphic in this species, adult females have deeper voices than adult males in their whoop and groan vocalizations (Theis et al. 2007; Mathevon et al. 2010), so these calls inform listeners about the sex of callers.

Female spotted hyenas are larger than males because they grow faster, rather than showing a longer period of growth (Swanson et al. 2013). Eleven sets of male and female littermates born as mixed sex twins at the Berkeley hyena colony did not differ in mass at birth; on average, both sexes weighed 1.5 kg at birth (Fig. 4A). However, there was a significant effect of litter size on mass at birth, with singletons being heaviest and members of triplet litters being lightest (Fig. 4B). Early in postnatal life males and females

appear to grow similarly, but between weaning and reproductive maturity their ontogenetic growth trajectories diverge. Female growth rates increase relative to those of males as animals approach sexual maturity. Traits that mature before divergence of these ontogenetic trajectories are monomorphic, whereas traits that mature later are dimorphic (Swanson et al. 2013). Although it is difficult to distinguish young males from young females, in adulthood, female spotted hyenas are visibly brawnier than their male peers (Fig. 5).

The teeth of a spotted hyena, particularly the incisors and canine teeth, are its primary weapons, but neither the canines nor the incisors differ significantly between the sexes in spotted hyenas (Van Horn et al. 2003). However, in adulthood, the height of the lower canine tooth tends to be slightly larger in females than males, so if any sexual dimorphism exists at all in weaponry in this species, females may have a slight advantage.

External Genitalia

The external genitalia of female spotted hyenas are unique among mammals. The clitoris is greatly elongated to form a fully erectile structure (Fig. 6), with a single urogenital (UG) tract passing from the tip of this structure into the caudal region of the abdomen (Matthews 1939; Neaves et al. 1980; Henschel and Skinner 1991). The female's enlarged clitoris is not as slender as the male's penis (Glickman et al. 2006), and the clitoris is slightly shorter (mean = 17 cm) than the male's penis (mean = 19 cm) (Neaves et al. 1980; Drea et al. 2002). Furthermore, the glans of the phallus is blunt and barrel-shaped in females, whereas it is angular and pointed in males (Fig. 7; Frank et al. 1990; Cunha et al. 2003). Nevertheless, the female's external genitalia look remarkably like those of the male (Figs. 5–7). In female spotted hyenas, these unusual genitalia are present at birth. "Masculinization" of their genitalia is not as strictly androgen-dependent (Drea et al. 1998; Cunha et al. 2005; Conley et al. 2020) as it is in other mammals. Initial prenatal development of the external genitalia in both sexes is

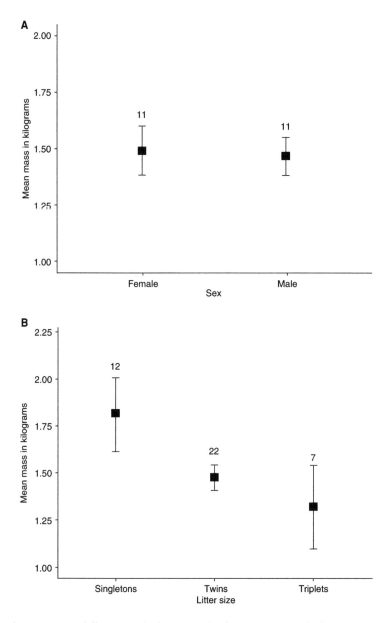

Figure 4. (*A*) There is no sex difference in body mass at birth among twin cubs born in mixed sex litters in captivity at the Berkeley hyena colony (Wilcox Rank Sum Test: $P = 0.553$); each litter contained one male and one female cub. (*B*) Litter size does have a significant effect on cub birthweights (Kruskal Wallace Test: $P = 0.0007$). Error bars represent 95% confidence intervals. Numbers above the error bars indicate individuals sampled.

largely free of androgenic influence (Glickman et al. 2006); however, elevated androgen concentrations in late gestation influence the development of both genital morphology and behavior (Drea et al. 2002; Dloniak et al. 2006; Holekamp et al. 2013). Furthermore, AA treatment of preg- nant females reduces the developmental influ- ence of androgens on their fetuses (Conley et al. 2020). No adult females in other members of the hyena family have unusual genitalia, so the enlargement of the clitoris is unique to spotted hyenas. In contrast to other carnivores, includ-

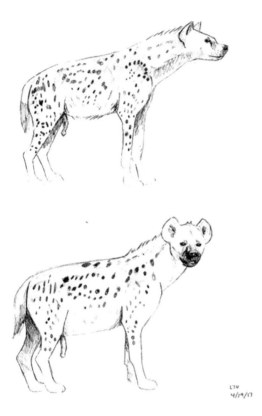

Figure 5. Adult male (*top*) and adult female hyena (*bottom*) showing sex differences in body shape and genital morphology. (Drawing by Lily Johnson-Ulrich.)

ing the other hyena species, the female spotted hyena has no external vaginal opening. Instead, she urinates, copulates, and gives birth through the tip of her elongated clitoris. Her vaginal labia are fused together and filled with fat and connective tissue to form a bilobed structure (Fig. 6) that resembles the scrotal sac and testes of the male (Frank et al. 1990).

Although the spotted hyena's clitoris and penis are similar in length, their internal anatomies are strikingly different to match their different functions (Cunha et al. 2005). The male urethra needs only to allow for the passage of urine and ejaculate, but in addition to passing urine, the female UG canal that traverses the clitoris must enable her to receive the male during copulation and to give birth to cubs that weigh ~1.5 kg. In contrast to the male urethra, which is narrow and surrounded by the corpus

spongiosum, the female UG canal is more pleated, voluminous, and expandable because it is surrounded by loose connective tissue, which facilitates the birthing process. The spotted hyena's penis and clitoris are retractable organs as a result of retractor muscles that span their lengths. However, the position of the retractor muscles relative to the urethra and the UG canal are quite different in the two sexes, with the retractor muscles being ventral to the urethra in males and dorsal to the UG canal in females. If not for this sex difference, the retractor muscles within the clitoris would surely be damaged during parturition, because the distal clitoris tears along its ventral midline (Fig. 6) during the first birth of a cub, which has a cranial diameter substantially exceeding the diameter of the clitoral meatus. Interestingly, the differences in the internal anatomies of the penis and clitoris are androgen dependent, as indicated by the ef-

Figure 6. External genitalia of the female spotted hyena seen while this female engages in a greeting ceremony with another female, who sniffs her anogenital region. The vaginal labia form structures that resemble the male's scrotal sac, and the clitoris is elongated to form a fully erectile "pseudopenis." The strip of pink scar tissue running down the posterior surface of the clitoris was caused by tearing during parturition, and indicates that the female has borne at least one litter in the past.

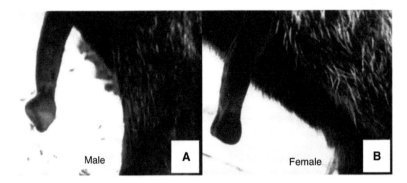

Figure 7. The gross anatomy of the male and female genitalia of the spotted hyena *(Crocuta crocuta)*. Shown are the penis (*A*), and clitoris (*B*) in the erect state.

fects of AAs that were administered to pregnant dams starting in early gestation. Most striking was the finding that the internal penile anatomy of AA-treated males was almost completely "feminized"—the urethra was more pleated and voluminous, because it was surrounded by loose connective tissue rather than the corpus spongiosum, and the retractor muscles had shifted from the ventral to dorsal position relative to the urethra (Cunha et al. 2005). Many of the external differences between the penis and clitoris (as described above) were also erased by in utero AA treatment, in that the penis was reduced in length, and the shape of the glans was feminized. This alteration proved to be functionally significant, because the males that had received the most intensive AA treatment during fetal development were incapable of copulating with receptive females (Drea et al. 2002).

In both sexes, the phallus is erect during greeting ceremonies, but during copulation, only the penis is erect while the clitoris is flaccid and retracted. The location of the clitoral opening is far more rostral than in more typical mammals, and as such, males perform penile "flips" that are used to locate the flaccid and retracted clitoris, and then pull the opening caudally so that he can intromit. The ability of an AA-treated male to reach the clitoral opening was largely negated by the reduced length of his feminized penis. And the AA treatment also feminized the perineal muscles that contribute to the flipping behavior (i.e., the bulbocavernosus [BC] muscles) (Forger et al. 1996). As a result, AA-treated

males are more likely to misdirect their flips (NJ Place and SE Glickman, unpubl.). Similar to the internal anatomy of the phallus, the morphology of the BC muscles is sexually dimorphic in spotted hyenas and in utero AA treatment feminized the BC muscles in males. The BC muscles are innervated by spinal motoneurons located in Onuf's nucleus (see below), and the male advantage in the number of motoneurons was negated in AA-treated males.

Nervous System

Sex differences in behavior are of course mediated by differences in the nervous system. One would therefore expect to see a mosaic pattern of typical and atypical sex differences in the spinal cords and brains of spotted hyenas, as we do in the behavior and morphology of these animals. Regions such as those mediating sexual behavior (e.g., male-typical mounting and erection of the phallus), aggression, and dispersal are of special interest in this regard. Although there are major obstacles impeding research on the hyena's nervous system, we are aware of five such studies. We begin with three that focused on specialized subpopulations of cells within the nervous system, and then turn to two that have taken a broader "whole brain" approach.

The first study to compare the nervous systems of male and female hyenas described spinal motor neurons located in Onuf's nucleus (Forger et al. 1996). These cells project to muscles at the base of the penis, and in most mammals, they are more numerous in males than females (Sakamoto 2014). This is also the case in spotted hyenas; adult males have ~20% more Onuf's neurons than do females. This difference is present at birth, and prenatal AA treatment feminizes the nucleus in males (Forger et al. 1996). These data might suggest that hyenas are not unusual with respect to Onuf's nucleus. However, both male and female hyenas frequently engage in greeting ceremonies in which the erect phallus of one animal is presented to another for sniffing (Fig. 6), so perhaps motor neurons in Onuf's nucleus play a role in this behavior in both sexes (Forger et al. 1996). As male hyenas face unusual challenges associated with copulation that females do not, these behaviors may depend on the additional neurons found in the male's Onuf's nucleus. This suggests that the development of the motor neurons projecting to muscles controlling the phallus in females follows the same "rules" as the development of the phallus itself; this in turn suggests considerable masculinization via androgen-independent mechanisms. The additional development of Onuf's nucleus in males may be mediated by androgens produced in the testes, which is the case for the phallus (Glickman et al. 2006).

The first subcortical region of the brain examined in spotted hyenas is the sexually dimorphic nucleus (SDN) in the preoptic area (Fenstemaker et al. 1999), which is larger in males than females in the many species in which it has been examined (McCarthy et al. 2017). This is also the case in spotted hyenas. In other species, the SDN is associated with mounting, intromission, and ejaculation, as well as partner preference (for review, see Pfaff and Baum 2018). Although the motor coordination of copulation may seem especially challenging for male hyenas, it can be decomposed into the same basic elements as seen in other species, and it is directed toward females. It is therefore not surprising that the sex difference in the SDN of hyenas would resemble that found in other mammals (i.e., that it is larger in males than in females). However, the two-fold sex difference in size of the hyena's SDN is considered modest compared with those in other species (Fenstemaker et al. 1999).

Rosen et al. (2006) assessed sex differences in four forebrain regions of captive hyenas with respect to the density of fibers containing the peptide vasopressin (VP). In many other species, VP innervation of the forebrain, particularly that of the lateral septum, is associated with social behaviors such as aggression and dominance (Albers 2012), and VP innervation of the lateral septum is consistently greater in males than in females (De Vries and Panzica 2006). Rosen et al. (2006) found no sex differences in the sub-paraventricular region of the hypothalamus, anterior hypothalamic region, or anterior supraoptic region, nor was there a significant sex difference in the lateral septum, but they observed a bimodal distribution of VP fibers in this brain region in males. Specifically, they found that VP fibers were heavily concentrated in the lateral septum of all three females examined and in two of the four males, but that these fibers were virtually absent in the other two males. These investigators suggested that in a natural setting, VP may contribute to the heightened aggression of adult females relative to adult males that have dispersed from their natal clans, but that their two captive males with elevated VP, which were both living with peers, may have been in a predispersal state, in which their aggressive behavior had not yet declined to facilitate immigration into a neighboring clan.

Sex differences have also been examined with respect to whole brains, endocranial volumes and "virtual brains" of spotted hyenas (Arsznov et al. 2010; Mann et al. 2018). Mann et al. (2018) reached the conclusion that "females have smaller brains despite having bodies that are (on average) longer and heavier," whereas Arsznov et al. (2010) found no evidence of a sex difference in overall brain size but did see regional sex differences. The methods used to collect and analyze data were quite different in these two studies so it is difficult to compare them.

Mann et al. (2018) found no sex difference in the weight of brains dissected from nine hyenas in Northern Kenya or in endocranial volumes in a collection of 60 skulls in the British Museum of Natural History (BMNH), but they did find a difference, favoring males, in a sample of 19 skulls in the Museum of Vertebrate Zoology (MVZ) at UC Berkeley. Body lengths were greater in females than males in the BMNH collec-

Table 1. A summary of traits reviewed here, indicating which conform to a typical mammalian pattern of sexual dimorphism, which show a "role-reversed" pattern of dimorphism, and which show no sex differences

Category	Typical mammalian pattern	Role-reversed pattern	No sex difference
Behavior	Space-use	Aggressive behavior	Cognition
	Dispersal behavior	Social dominance	Hunting
	Sexual behavior	Territory defense	
	Parental behavior		
Morphology	Onuf's nucleus	Body size	Weaponry
	SDN		

(SDN) Sexually dimorphic nucleus of the hypothalamus.

tion, but there was no difference in the MVZ collection. These patterns thus suggest a difference favoring males in the BMNH sample because body size was greater in females, and in the MVZ sample because brain size was smaller in females. Reasons for these interpopulation differences are not obvious. In contrast, Arsznov et al. (2010) found no evidence of a sex difference in overall brain sizes measured from a sample of 22 adult hyena skulls in the Michigan State University Museum. Here, brain volumes were measured in "virtual" brains (endocasts) created from CAT scans of skulls; these volumes were divided by skull basal length to take body size into account. There were many differences between the protocols used by Mann et al. (2018) and Arsznov et al. (2010) that might account for what appear to be conflicting results. For example, the former study evaluated sex differences in brain size and in body length but not in brain size corrected for body length, whereas investigators of the latter study used skull length as a proxy for body size and conducted their analysis on the corrected values.

When Arsznov et al. (2010) examined regional brain volumes (as a proportion of total brain volume) in virtual hyena brains, they found clear and interesting differences between the sexes. Specifically, the anterior cerebrum was larger in males than females, and the posterior cerebrum was larger in females than males. The latter difference is difficult to interpret because this portion of the brain contains a multitude of subregions that have a diverse array of functions but that were impossible to delineate with the endocast method. The anterior cerebrum contains primarily frontal cortex, which is associat-

ed with a variety of measures of social cognition, as well as inhibitory control. However, no differences between male and female hyenas have been found to date in either of these domains in behavioral tests (Benson-Amram and Holekamp 2012; Johnson-Ulrich and Holekamp 2020).

CONCLUSIONS

In the domains of both behavior and morphology, we find in spotted hyenas a mosaic of traits: some conform to sex differences in other mammals, whereas others do not (Table 1). We find that spotted hyenas show sexually dimorphic behavior that conforms to mammalian norms with respect to space-use, dispersal behavior, sexual behavior, and parental behavior. However, we also find sex differences that are reversed from mammalian norms in the hyena's aggressive behavior, social dominance, and territory defense. Neither weapons nor hunting behavior differ between males and females, but sexual size dimorphism is distinctly reversed from mammalian norms. However, as in other mammals, the SDN in spotted hyenas is larger in males than in females (Fenstemaker et al. 1999). Similarly, the number of motoneurons innervating the perineal muscles associated with the phallus is sexually dimorphic in the conventional manner (Forger et al. 1996). Several features of VP immunoreactivity in the spotted hyena forebrain are similar to what has been described in other mammals (Rosen et al. 2006). However, contrary to what has been reported in many other species (De Vries and Panzica 2006), the density of VP innervation of the lateral septum is not sexually dimorphic. Whether there are sex differences in overall brain size remains controversial. In any case, it is clear that "sex role-reversed" or "masculinized" traits coexist in the same females with other sexually dimorphic traits that are like those shown by virtually all "typical" female mammals.

The female spotted hyena's chimeric blend of feminine and "masculinized" traits raises questions about how and why some traits have been "reversed" in this species, whereas others have not. Overall it appears that the behavioral traits that have been "masculinized" in female spotted hyenas are those critical to ensuring consistent access to food resources, which is the critical factor limiting reproductive success in females of this species (Holekamp and Strauss 2020). Aggressive behavior, social dominance, and territory defense are all very important in this regard. Furthermore, because body size is not a good predictor of fight outcomes among spotted hyenas (Smale et al. 1993), larger body size does not help female hyenas win fights; instead, larger body size most likely helps females capture larger prey to help satisfy the enormous energetic demands imposed by pregnancy and lactation. Evidently, natural selection has acted to modify those traits related to food access from the ancestral condition, but has left intact those behavioral and morphological traits that are unrelated to accessing food, such that they conform to patterns of sexual dimorphism in other mammals. The adaptive significance of the female's odd genitalia remains uncertain, although they may play a role in allowing females to determine which sperm fertilize her ova.

ACKNOWLEDGMENTS

First and foremost, we thank our friend and colleague Steve Glickman, who passed away before this paper was finished. In one way or another, Steve introduced all of us to the joys of studying what he recognized to be the coolest animal ever, the spotted hyena. Steve developed and maintained the captive hyena colony at the University of California, Berkeley, from the mid-1980s until 2010. One of Steve's original goals was to elucidate sex differences in this peculiar species, and his many years of work on captive hyenas made many advances in this area. We only wish he had lived to see this paper published.

We also thank the Kenyan National Commission for Science, Technology, and Innovation, the Narok County Government, the Mara Conservancy, and the Kenya Wildlife Service for permission to conduct our long-term study. We also thank all those who assisted with data collection in the field, and with data entry, and manipulation in the laboratory. This work was supported by National Science Foundation (NSF) Grants OISE1853934 and IOS1755089 to K.E.H. S.K.M. was supported by OISE18 53934, a University Distinguished Fellowship from Michigan State University and a Graduate Research Fellowship from NSF.

REFERENCES

Albers E. 2012. The regulation of social recognition, social communication and aggression: vasopressin in the social behavior neural network. *Horm Behav* **61**: 283–292. doi:10.1016/j.yhbeh.2011.10.007

Arsznov BM, Lundrigan BL, Holekamp KE, Sakai ST. 2010. Sex and the frontal cortex: a developmental CT study in the spotted hyena. *Brain Behav Evol* **76**: 185–197. doi:10.1159/000321317

Beach FA, Buehler MG, Dunbar IF. 1982. Competitive behavior in male, female and pseudo-hermaphroditic female dogs. *J Comp Physiol Psych* **96**: 855–874. doi:10.1037/0735-7036.96.6.855

Benhaiem S, Hofer H, Kramer-Schadt S, Brunner E, East ML. 2012. Sibling rivalry: training effects, emergence of dominance and incomplete control. *Proc Biol Sci* **279**: 3727–3735. doi:10.1098/rspb.2012.0925

Benson-Amram SR, Holekamp KE. 2012. Innovative problem solving by wild spotted hyenas. *Proc Biol Sci* **279**: 4087–4095. doi:10.1098/rspb.2012.1450

Boydston EE, Morelli TL, Holekamp KE. 2001. Sex differences in territorial behavior exhibited by the spotted hyena (Hyaenidae, *Crocuta crocuta*). *Ethology* **107**: 369–385. doi:10.1046/j.1439-0310.2001.00672.x

Boydston EE, Kapheim KM, Van Horn RC, Smale L, Holekamp KE. 2005. Sexually dimorphic patterns of space use throughout ontogeny in the spotted hyena (*Crocuta crocuta*). *J Zool Lond* **267**: 271–281. doi:10.1017/S0952 836905007478

Buchan JC, Alberts SC, Silk JB, Altmann J. 2003. True paternal care in a multi-male primate society. *Nature* **425**: 179–181. doi:10.1038/nature01866

Conley A, Place NJ, Legacki EL, Hammond GL, Cunha GR, Drea CM, Weldele ML, Glickman SE. 2020. Spotted hyaenas and the sexual spectrum: reproductive endocrinology and development. *J Endocrinol* **247**: R27–R44. doi:10.1530/JOE-20-0252

Cunha GR, Wang Y, Place NJ, Liu W, Baskin L, Glickman SE. 2003. Urogenital system of the spotted hyena (*Crocuta crocuta* Erxleben): a functional histological study. *J Morph* **256:** 205–218. doi:10.1002/jmor.10085

Cunha GR, Place NJ, Baskin L, Conley A, Weldele ML, Cunha TJ, Wang YZ, Cao M, Glickman SE. 2005. The ontogeny of the urogenital system of the spotted hyena (*Crocuta crocuta* Erxleben). *Biol Reprod* **73:** 554–564. doi:10.1095/biolreprod.105.041129

De Vries GJ, Panzica GC. 2006. Sexual differentiation of central vasopressin and vasotocin systems in vertebrates: different mechanisms, similar endpoints. *Neuroscience* **138:** 947–955. doi:10.1016/j.neuroscience.2005.07.050

Dloniak SM, French JA, Holekamp KE. 2006. Rank-related maternal effects of androgens on behaviour in wild spotted hyenas. *Nature* **440:** 1190–1193. doi:10.1038/nature04540

Drea CM, Weldele ML, Forger NG, Coscia EM, Frank LG, Licht P, Glickman SE. 1998. Androgens and masculinization of genitalia in the spotted hyena (*Crocuta crocuta*). 2: Effects of prenatal anti-androgens. *J Reprod Fertil* **113:** 117–127. doi:10.1530/jrf.0.1130117

Drea CM, Place NJ, Weldele ML, Coscia EM, Licht P, Glickman SE. 2002. Exposure to naturally circulating androgens during foetal life incurs direct reproductive costs in female spotted hyenas, but is prerequisite for male mating. *Proc Biol Sci* **269:** 1981–1987. doi:10.1098/rspb.2002.2109

East ML, Hofer H. 2001. Male spotted hyenas (*Crocuta crocuta*) queue for status in social groups dominated by females. *Behav Ecol* **12:** 558–568. doi:10.1093/beheco/12.5.558

Fenstemaker SB, Zup SL, Frank LG, Glickman SE, Forger NG. 1999. A sex difference in the hypothalamus of the spotted hyena. *Nat Neurosci* **2:** 943–945. doi:10.1038/14728

Forger NG, Frank LG, Breedlove SM, Glickman SE. 1996. Sexual dimorphism of perineal muscles and motoneurons in spotted hyenas. *J Comp Neurol* **375:** 333–343. doi:10.1002/(SICI)1096-9861(19961111)375:2<333::AID-CNE11>3.0.CO;2-W

Frank LG, Glickman SE. 1994. Giving birth through a penile clitoris: parturition and dystocia in the spotted hyaena (*Crocuta crocuta*). *J Zool Lond* **234:** 659–665. doi:10.1111/j.1469-7998.1994.tb04871.x

Frank LG, Glickman SE, Zabel CJ. 1989. Ontogeny of female dominance in the spotted hyaena: perspectives from nature and captivity. In *The biology of large African mammals in their environment (Symposium of the Zoological Society of London)* (ed. Maloiy G and Jewell P), pp. 127–146. Oxford University Press, Oxford.

Frank LG, Glickman SE, Powch I. 1990. Sexual dimorphism in the spotted hyaena (*Crocuta crocuta*). *J Zool Lond* **221:** 308–313. doi:10.1111/j.1469-7998.1990.tb04001.x

Frank LG, Glickman SE, Licht P. 1991. Fatal sibling aggression, precocial development, and androgens in neonatal spotted hyaenas. *Science* **252:** 702–704. doi:10.1126/science.2024122

Glickman SE, Cunha GR, Drea CM, Conley AJ, Place NJ. 2006. Mammalian sexual differentiation: lessons from the spotted hyena. *Trends Endocrinol Metab* **17:** 349–356. doi:10.1016/j.tem.2006.09.005

Golla W, Hofer H, East ML. 1999. Within-litter sibling aggression in spotted hyaenas: effect of maternal nursing, sex and age. *Anim Behav* **58:** 715–726. doi:10.1006/anbe.1999.1189

Henschel JR, Skinner JD. 1991. Territorial behaviour by a clan of spotted hyaenas *Crocuta crocuta*. *Ethology* **88:** 223–235. doi:10.1111/j.1439-0310.1991.tb00277.x

Hofer H, East ML. 1995. Population dynamics, population size, and the commuting system of Serengeti spotted hyaenas. In *Serengeti II: dynamics, management, and conservation of an ecosystem* (ed. Sinclair ARE, Arcese P), pp. 332–363. University of Chicago Press, Chicago.

Hofer H, East ML. 1996. The components of parental care and their fitness consequences: a life history perspective. *Verh Dtsch Ges Zool Ges.* **89.2:** 149–164.

Hofer H, Benhaiem S, Golla W, East ML. 2016. Trade-offs in lactation and milk intake by competing siblings in a fluctuating environment. *Behav Ecol* **27:** 1567–1578. doi:10.1093/beheco/arw078

Holekamp KE. 1984. Dispersal in ground-dwelling sciurids. In *The biology of ground-dwelling squirrels* (ed. Murie JO, Michener GR), pp. 297–320. University of Nebraska Press, Lincoln, NE.

Holekamp KE, Strauss ED. 2020. Reproduction within a hierarchical society from a female's perspective. *Integr Comp Biol* **60:** 753–764. doi:10.1093/icb/icaa068

Holekamp KE, Smale L, Berg R, Cooper SM. 1997. Hunting rates and hunting success in the spotted hyena (*Crocuta crocuta*). *J Zool Lond* **242:** 1–15. doi:10.1111/j.1469-7998.1997.tb02925.x

Holekamp KE, Van Meter PE, Swanson EM. 2013. Developmental constraints on behavioural flexibility. *Philos Trans R Soc B Biol Sci* **368:** 20120350. doi:10.1098/rstb.2012.0350

Höner OP, Wachter B, East ML, Streich WJ, Wilhelm K, Burke T, Hofer H. 2007. Female mate-choice drives the evolution of male-biased dispersal in a social mammal. *Nature* **448:** 798–801. doi:10.1038/nature06040

Johnson-Ulrich L, Holekamp KE. 2020. Group size and social rank predict inhibitory control in spotted hyaenas. *Anim Behav* **160:** 157–168. doi:10.1016/j.anbehav.2019.11.020

Kolowski JM, Katan D, Theis KR, Holekamp KE. 2007. Daily patterns of activity in the spotted hyena. *J Mammal* **88:** 1017–1028. doi:10.1644/06-MAMM-A-143R.1

Kruuk H. 1972. *The spotted hyena.* University of Chicago Press, Chicago.

Lincoln GA. 1994. Teeth horns and antlers: the weapons of sex. In *The differences between the sexes* (ed. Short RV, Balaban E), pp. 131–158. Cambridge University Press, Cambridge.

Mann MD, Frank LG, Glickman SE, Towe AL. 2018. Brain and body size relations among spotted hyenas (*Crocuta crocuta*). *Brain Behav Evol* **92:** 82–95. doi:10.1159/000494125

Mathevon N, Koralek N, Weldele M, Glickman SE, Theunissen FE. 2010. What the hyena's laugh tells: sex, age, dominance and individual signature in the giggling call of *Crocuta crocuta*. *BMC Ecol* **10:** 9. doi:10.1186/1472-6785-10-9

Cite this article as *Cold Spring Harb Perspect Biol* doi: 10.1101/cshperspect.a039180

Matthews LH. 1939. Reproduction in the spotted hyaena, *Crocuta crocuta* (Erxleben). *Phil Trans R Soc Lond B* **230:** 1–78. doi:10.1098/rstb.1939.0004

McCarthy MM, De Vries GJ, Forger NG. 2017. Sexual differentiation of the brain: a fresh look at mode, mechanisms, and meaning. In *Hormones, brain, and behavior*, 3rd ed. (ed. Pfaff DW, Joëls M), Vol. 5, pp. 3–32. Academic, Orlando, FL.

McCormick SK, Holekamp KE. Aggressiveness and submissiveness in spotted hyenas. *Anim Behav* (in review).

Mills MGL. 1990. *Kalahari hyaenas*. Unwin-Hyman, London.

Neaves WB, Griffin JE, Wilson JD. 1980. Sexual dimorphism of the phallus in spotted hyaena (*Crocuta crocuta*). *J Reprod Fert* **59:** 509–513. doi:10.1530/jrf.0.0590509

Oftedal OT, Gittleman JG. 1989. Patterns of energy output during reproduction in carnivores. In *Carnivore behavior, ecology and evolution* (ed. Gittleman JG), pp. 355–378. Cornell University Press, Ithaca, NY.

Pfaff DW, Baum MJ. 2018. Hormone-dependent medial preoptic/lumbar spinal cord/autonomic coordination supporting male sexual behaviors. *Mol Cell Endocrinol* **467:** 21–30. doi:10.1016/j.mce.2017.10.018

Pusey AE, Packer C. 1987. Dispersal and philopatry. In *Primate societies* (ed. Smuts BB, Cheney DL, Seyfarth RM, Wrangham RW, Struhsaker TT), pp. 250–266. University of Chicago Press, Chicago.

Rosen GJ, De Vries GJ, Villalba C, Weldele ML, Place NJ, Coscia EM, Glickman SE, Forger NG. 2006. The distribution of vasopressin in the forebrain of spotted hyenas. *J Comp Neurol* **498:** 80–92. doi:10.1002/cne.21032

Sakamoto H. 2014. Sexually dimorphic nuclei in the spinal cord control male sexual functions. *Front Neurosci* **8:** 1–6.

Smale L, Frank LG, Holekamp KE. 1993. Ontogeny of dominance in free-living spotted hyaenas: juvenile rank relations with adults females and immigrant males. *Anim Behav* **46:** 467–477. doi:10.1006/anbe.1993.1215

Smale L, Holekamp KE, Weldele M, Frank LG, Glickman SE. 1995. Competition and cooperation between litter-mates in the spotted hyena, *Crocuta crocuta. Anim Behav* **50:** 671–682. doi:10.1016/0003-3472(95)80128-6

Smale L, Nunes S, Holekamp KE. 1997. Sexually dimorphic dispersal in mammals: patterns, causes, and consequences. *Adv Study Behav* **26:** 181–250. doi:10.1016/S0065-3454(08)60380-0

Smale L, Holekamp KE, White PA. 1999. Siblicide revisited in the spotted hyaena: does it conform to obligate or facultative models? *Anim Behav* **58:** 545–551. doi:10.1006/anbe.1999.1207

Strauss ED, Holekamp KE. 2019. Social alliances improve rank and fitness in convention-based societies. *Proc Nat Acad Sci* **116:** 8919–8924. doi:10.1073/pnas.1810384116

Swanson EM, Dworkin I, Holekamp KE. 2011. Lifetime selection on a hypoallometric size trait in the spotted hyena. *Proc Biol Sci* **278:** 3277–3285. doi:10.1098/rspb.2010.2512

Swanson EM, McElhinny TL, Dworkin I, Weldele ML, Glickman SE, Holekamp KE. 2013. Ontogeny of sexual size dimorphism in the spotted hyena (*Crocuta crocuta*). *J Mamm* **94:** 1298–1310. doi:10.1644/12-MAMM-A-277.1

Szykman M, Engh AL, Van Horn RC, Boydston EE, Scribner KT, Holekamp KE. 2003. Rare male aggression directed toward females in a female-dominated society: baiting behavior in the spotted hyena. *Aggress Behav* **29:** 457–474. doi:10.1002/ab.10065

Szykman M, Van Horn RC, Engh AL, Boydston EE, Holekamp KE. 2007. Courtship and mating in free-living spotted hyenas. *Behaviour* **144:** 815–846. doi:10.1163/156853907781476418

Theis KR, Greene KM, Benson-Amram SR, Holekamp KE. 2007. Sources of variation in the long-distance vocalizations of spotted hyenas. *Behaviour* **144:** 557–584. doi:10.1163/156853907780713046

Van Horn RC, McElhinny TL, Holekamp KE. 2003. Age estimation and dispersal in the spotted hyena (*Crocuta crocuta*). *J Mamm* **84:** 1019–1030. doi:10.1644/BBa-023

Vullioud C, Davidian E, Wachter B, Rousset F, Courtiol A, Höner OP. 2019. Social support drives female dominance in the spotted hyaena. *Nat Ecol Evol* **3:** 71–76. doi:10.1038/s41559-018-0718-9

Wachter B, Höner OP, East ML, Golla W, Hofer H. 2002. Low aggression levels and unbiased sex ratios in a prey-rich environment: no evidence of siblicide in Ngorongoro spotted hyenas (*Crocuta crocuta*). *Behav Ecol Sociobiol* **52:** 348–356. doi:10.1007/s00265-002-0522-y

Wahaj SA, Holekamp KE. 2006. Functions of sibling aggression in the spotted hyena (*Crocuta crocuta*). *Anim Behav* **71:** 1401–1409. doi:10.1016/j.anbehav.2005.11.011

Watts HE, Tanner JB, Lundrigan BL, Holekamp KE. 2009. Post-weaning maternal effects and the evolution of female dominance in the spotted hyena. *Proc Biol Sci* **276:** 2291–2298.

Yoshida KCS, Van Meter PE, Holekamp KE. 2016. Variation among free-living spotted hyenas in three personality traits. *Behaviour* **153:** 1665–1722. doi:10.1163/1568539X-00003367

Evidence for Perinatal Steroid Influence on Human Sexual Orientation and Gendered Behavior

Ashlyn Swift-Gallant,[1] Talia Shirazi,[2] David A. Puts,[2] and S. Marc Breedlove[3]

[1]Department of Psychology, Memorial University of Newfoundland, St. John's A1B 3X9, Newfoundland and Labrador

[2]Department of Anthropology, The Pennsylvania State University, University Park, Pennsylvania 16802, USA

[3]Neuroscience Program, Michigan State University, East Lansing, Michigan 48824, USA

Correspondence: breedsm@msu.edu

In laboratory animals, exposure to gonadal steroid hormones before and immediately after birth can exert permanent effects on many behaviors, particularly reproductive behaviors. The extent to which such effects occur in humans remains an open question, but several lines of evidence indicate that perinatal levels of both androgens and estrogens may affect adult human psychology and behavior, including sexual orientation and gender nonconformity. Some putative indicators of prenatal androgen exposure, including the ratio of the length of the index finger to that of the ring finger (2D:4D), have repeatedly indicated that lesbians, on average, were exposed to more prenatal androgens than straight women, suggesting that sufficient fetal androgen exposure predisposes a fetus to gynephilia (attraction to women) at maturity. The digit ratios of gay men do not differ from those of straight men, suggesting that prenatal androgen levels are not responsible for their androphilia (attraction to men). However, evidence that gay men who prefer an insertive anal sex role (ASR) have more masculine digit ratios than those preferring a receptive ASR suggests that early androgens influence some sexual preferences in men. Furthermore, digit ratios among gay men have been found to correlate with recalled childhood gender nonconformity (CGN). People with isolated gonadotropin-releasing hormone (GnRH) deficiency (IGD) offer further insight into the effects of perinatal gonadal steroid exposure. In people with IGD, gonadal hormone production is low or absent after the first trimester of gestation. However, because placental gonadotropins drive gonadal hormone secretion during the first trimester when genitalia sexually differentiate, individuals with IGD are unambiguously male or female at birth, consistent with their chromosomal and gonadal sex. Men with IGD report greater CGN, again suggesting that perinatal androgen exposure contributes to male-typical behavioral patterns in humans. Interestingly, women with IGD report less androphilia and more bisexuality than control women, suggesting that perinatal ovarian steroids in females typically augment androphilia in adulthood. Taken together, these findings indicate that the perinatal hormonal milieu influences human sexual orientation and gender conformity.

The ongoing question of whether and how perinatal steroid hormones, including androgens such as testosterone and estrogens such as estradiol, have a lasting influence on human behavior is not likely to be settled soon. The question arises because of the voluminous studies in mammalian models, where exposure to testosterone and/or its metabolites, both androgens and estrogens, permanently alters the morphology of the brain and spinal cord, as well as virtually every behavior that normally differs between the sexes (Phoenix et al. 1959; McCarthy et al. 2012). At the least, we can speculate that any influence of perinatal hormones in humans may generally be less robust than in animal models, because several clinical cases of atypical perinatal hormone exposure, such as congenital adrenal hyperplasia (CAH), might otherwise have very pronounced effects on behavior, yet studies of women with CAH report modest effects on juvenile play, sexual orientation, and other behaviors (Spencer et al. 2021). Even those effects of CAH may occur in part due to social factors, including parental attention and concerns, medical management, and/or the awareness of effects of the condition on genitalia, rather than any permanent effects that hormones might work directly upon the brain. Similarly, the feminine behavior of women with androgen insensitivity syndrome (AIS) (Hines 2011), a result of disrupted androgen receptor signaling, could be caused either by the absence of androgen effects on the brain, elevated estrogen levels relative to unaffected XY individuals (Hughes and Deeb 2006), and/or their social upbringing as females.

These issues have inspired us to seek evidence of perinatal influences on human sexual behaviors in circumstances where differences in social experience are less likely to provide a confounding variable to cloud interpretations. What, if any, effects can we discern on behavior when more subtle hormonal variations, which are not likely to be evident to the participant or their social sphere, are examined?

Although research exploring perinatal or "biological" influences on sex differences in human behavior may be regarded by some as an attempt to prove or justify viewing gender as a strict dichotomy—that every person should be regarded as either male or female, rather than a continuum—to our way of thinking, this is a false coupling of ideas. Rather than regarding perinatal hormonal influences as channeling each person down one of only two pathways, we regard variation in perinatal hormones, at least within each sex, as a potential source of additional variation in human behaviors, including gendered behaviors. Surely the wellspring of diversity in human behavior is due to the fact that it is extremely sensitive to so many diverse influences, including the happenstance of social experience and cultural milieu to be sure, but also the happenstance of the hormonal environment during development. Thus, evidence of hormonal effects could suggest greater diversity in human behavior, not less.

To illustrate this point, we will consider an adult marker that reflects prenatal androgens, digit ratios, and an endocrine condition, isolated gonadotropin-releasing hormone (GnRH) deficiency (IGD), that does not become evident until adolescence or young adulthood. In both cases, we are most interested in hormonal variation within each sex, where the typical genitalia at birth should provide at least grossly equivalent social influences through sex of rearing.

DIGIT RATIOS

Evidence that Digit Ratios Reflect Prenatal Androgen Exposure in Humans

Only four proximate mechanisms for producing a sex difference in the mammalian body have been documented to date. The sex chromosomes act directly in gonads to guide their differentiation. Testicular secretion of anti-Müllerian hormone (AMH) suppresses Müllerian duct development. Androgens such as testosterone and its metabolites masculinize the Wolffian ducts and external genitalia, and at puberty both testicular and ovarian hormones promote secondary sexual characteristics. Finally, social experience molds the sexes to dress and conduct themselves in accordance with the social norms around them. Thus, any morphological sex difference in the

Cite this article as *Cold Spring Harb Perspect Biol* doi: 10.1101/cshperspect.a039123

human body is likely due to at least one of these four mechanisms.

In 1998, John Manning reported that the ratio of the length of the second digit (2D, the index finger), divided by the length of the fourth digit (4D, the ring finger) was, on average, lower in human males than females across a wide range of ages, including 2-year-olds (Manning et al. 1998). The later reports that sex differences in 2D:4D are present in humans by the 14th week of gestation (Malas et al. 2006; Galis et al. 2010) rule out the possibility that this sex difference could originate from social influences.

That women with AIS, caused by dysfunction of the androgen receptor, have feminine digit ratios (Berenbaum et al. 2009; van Hemmen et al. 2017) directly implicates androgen signaling in the establishment of sex differences in digit ratios. Furthermore, the feminine nature of digit ratios in women with AIS also eliminates the possibility that either AMH or sex chromosomes establish this sex difference. These women carry a Y chromosome like males, and therefore develop testes that produce AMH, which suppresses their Müllerian duct development. If either AMH or sex chromosomes were responsible for masculinizing 2D:4D, then women with AIS should have masculine ratios. Thus, by elimination, the sex difference in digit ratios that cannot be attributed to the direct effect of sex chromosomes, AMH, or social influences, probably reflects the perinatal influence of androgens. Although other factors surely also influence digit ratios, including genes and estrogens, their feminine nature in women with AIS proves that functional androgen receptors are required for their full masculinization. Barring the discovery of a fifth mechanism guiding sexual differentiation of the mammalian body, we must conclude that the sex difference in human digit ratios is engendered by perinatal androgens.

Fortunately, a range of divergent findings corroborates the conclusion that prenatal and/ or neonatal androgens affect human digit ratios. People with CAH have more masculine (i.e., lower) digit ratios on the right hand than same-sex controls (Brown et al. 2002b; Ökten et al. 2002; Hönekopp and Watson 2010; Oswiecimska et al. 2012; Rivas et al. 2014). CAH has

little or no effect on the left-hand 2D:4D (Brown et al. 2002b; Buck et al. 2003; Nave et al. 2021), but then the sex difference in digit ratios is reduced on the left (Hönekopp and Watson 2010). For some reason, the right hand is seemingly more responsive to prenatal androgen than the left. Digit ratios are larger (more feminine) in men with Klinefelter syndrome (Manning et al. 2013; Chang et al. 2015), who experience reduced prenatal androgen secretions. In a nonclinical sample of women, digit ratios correlated with a firmly established readout of prenatal androgens, the anogenital distance, in the expected direction (Barrett et al. 2015). Anogenital distance also correlated with digit ratios in the expected direction in a clinical sample of prepubertal boys, where digit ratios also correlated with the extent of hypospadias (O'Kelly et al. 2020), a condition caused by reduced prenatal androgen stimulation in males.

Studies of digit ratios in nonhuman species generally support the idea that they reflect prenatal androgens, but whether 2D:4D is higher in females or males differs across species. Still, 2D:4D was masculinized (reduced) by prenatal androgen treatment in mice (Zheng and Cohn 2011), and correlated with prenatal androgens in the expected direction (smaller 2D:4D with greater androgen) in titi monkeys (Baxter et al. 2020), to cite two examples.

Because there is a sex difference in digit ratios that is masculinized in CAH, demasculinized in Klinefelter syndrome, and feminine in women with AIS, it would seem an inescapable conclusion that 2D:4D reflects perinatal androgen action in humans. Several researchers continue to express doubts about whether digit ratios actually do so (Leslie 2019), but no viable alternative hypothesis has been offered to account for the converging findings reviewed above.

Sexual Orientation in Women

One of the earliest findings of a relationship between digit ratios and human behavior was the 2000 *Nature* report that lesbians, on average, have a more masculine (smaller) digit ratio than straight women (Williams et al. 2000). This difference was seen only on the right hand, and this sample displayed a sex difference in digit ratios among straight people only on the right (Fig. 1), conforming to the conclusion of the meta-analyses noted above that the right hand is more responsive to prenatal androgen than the left. The difference in digit ratios between lesbians and straight women has been replicated numerous times (McFadden and Shubel 2002; Tortorice 2002; Hall and Love 2003; Rahman and Wilson 2003; Putz et al. 2004; Rahman 2005; Kraemer et al. 2006). A few researchers reported no such difference (Lippa 2003), but a meta-analysis of 16 data sets, including all that had been published at that time (Grimbos et al. 2010), concluded the difference was reliable, noting that more than 50 consecutive reports of no difference would be needed to render the effect size statistically nonsignificant. There have been several additional replications since, including reports that among monozygotic female twins discordant for sexual orientation, the lesbian twins have, on average, more masculine digit ratios than their sisters (Hiraishi et al. 2012; Watts et al. 2018), replicating an earlier report (Hall and Love 2003).

When participants were asked whether they considered themselves "butch" (more masculine) or "femme" (more feminine), two studies have found that lesbians who self-identified as "butch" had more masculine digit ratios than those who regarded themselves as "femme"

(Fig. 2; Brown et al. 2002a; Tortorice 2002). These findings complement reports that girls who experience gender dysphoria have more masculinized ratios than control girls (Wallien et al. 2008), and that women who regarded themselves as having been "tomboys" have lower 2D:4Ds than controls (Atkinson et al. 2017).

Given the converging evidence that prenatal androgens permanently affect digit ratios, the most parsimonious interpretation of these findings would be that lesbians, on average, were exposed to greater levels of prenatal androgen than straight women, which indicates that greater exposure to androgens before birth increases the likelihood that a person will be gynephilic (i.e., sexually attracted to women) upon reaching maturity. The differences between butch and femme lesbians suggests that at least some femme lesbians developed gynephilic interests for reasons unrelated to prenatal androgen exposure. Given that such women were surely among the lesbians in previous studies, any differences in digit ratios seen between straight and gay women would have had to overcome the dilution of an effect from inclusion of femme lesbians.

Sexual Orientation in Men

In contrast to the many reports of a difference in digit ratios in straight versus gay women, the

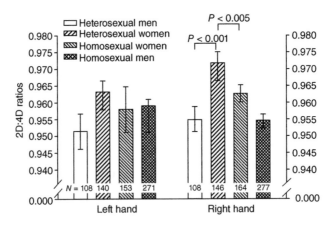

Figure 1. Sexual orientation and digit ratios. In this sample, a significant sex difference in digit ratios was seen only for the right hand. Women who reported a lesbian orientation had more masculine (smaller) right-hand digit ratios than self-reported straight women. No differences in digit ratios between gay and straight men were seen. (Figure reprinted from Williams et al. 2000 with permission from the authors.)

Cite this article as *Cold Spring Harb Perspect Biol* doi: 10.1101/cshperspect.a039123

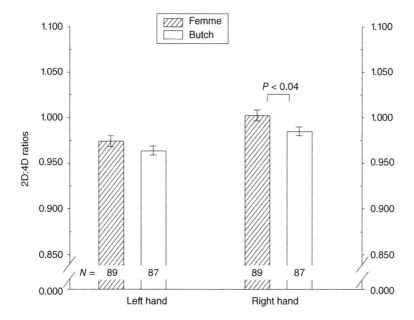

Figure 2. Digit ratios in butch and femme lesbians. Lesbians who self-report as being "butch" have more masculine (smaller) digit ratios than self-reported "femme" lesbians, but only on the right hand. (Figure reprinted from Brown et al. 2002a with permission from Springer © 2002.)

2000 report saw no differences in digit ratio between straight and gay men. In the years following, some researchers found gay men to have a less masculine digit ratio than straight men (Lippa 2003), but the reports were inconsistent (McFadden et al. 2005), and some found gay men to have more masculine digit ratios than straight men (Rahman and Wilson 2003). The 2010 meta-analysis integrating these findings concluded that gay and straight men do not differ overall in terms of digit ratios. This conclusion jibes well with reports using another presumptive marker of prenatal androgens, otoacoustic emissions. The sex difference in otoacoustic emissions, which is present at birth in humans, has been seen in other species, including sheep, in which prenatal testosterone treatment masculinized the measure (McFadden et al. 2009), confirming its sensitivity to androgens. Like digit ratios, otoacoustic emissions indicated that lesbians were exposed to more prenatal androgen than straight women, but that gay and straight men did not differ in prenatal androgens (McFadden and Pasanen 1998).

These findings indicate that variation in prenatal androgens is not responsible for the fact that gay men are androphilic. How can we reconcile the hypothesis that higher levels of prenatal androgen stimulation predisposes girls to be gynephilic in adulthood, versus the absence of a difference in prenatal androgen between gay and straight men? For one thing, it is possible that the reason the vast majority of men are gynephilic is because of their exposure to high levels of androgen before birth. There is little doubt that gay men were exposed to much higher levels of prenatal androgen than lesbians during some points in development (e.g., during genital development). Therefore, androphilia in gay men, if it is related to androgens at all, would be due either to a difference in their response to the abundant androgen they were exposed to before birth or a temporary difference in androgen exposure at some point in development other than the time(s) that digit ratios respond to the hormone.

It is also possible that the lack of a difference between gay and straight men in digit ratios is because, since there are surely multiple developmental pathways to a gay orientation, if either lower or higher levels of androgen predispose boys to be gay, lumping together these two populations would obscure either effect (Skorska and Bogaert 2017). If so, then the variance in reports of digit ratios in gay men, some finding gay men to have more masculine digit ratios (Rahman and Wilson 2003), others finding they have less masculine digit ratios (Robinson and Manning 2000), and yet others finding no difference (Williams et al. 2000), could be due to sampling differences in the proportion of the two types of gay men studied.

In an attempt to distinguish different types of gay men, participants were asked to report their preferred anal sex role (ASR), as either preferring an insertive role ("tops"), a receptive role ("bottoms"), or both roles ("versatiles"). As in previous studies, differences in digit ratios were found only for the right hand, where bottoms had larger (more feminine) digit ratios than tops. Versatiles had an intermediate digit ratio, not significantly different from either of the other two groups (Fig. 3). This result suggests that those males who grew up to be gay men preferring a receptive ASR were exposed to less prenatal androgen than gay men who prefer an insertive ASR.

Furthermore, conforming to previous reports (Moskowitz and Hart 2011; Zheng et al. 2012; Swift-Gallant et al. 2018), gay men who reported preferring a bottom role also reported more gender nonconformity, both in adulthood and growing up, than gay men preferring a top role. To see whether this difference in gender nonconformity might also be related to prenatal androgen levels, Swift-Gallant et al. tested for a relationship between right-hand digit ratios and gender nonconformity, both in adulthood and in childhood recollections, finding a positive correlation (Fig. 4). Taken together, these studies suggest that variation in prenatal androgen stimulation does affect the behavior of males, such that lower prenatal androgen stimulation predisposes males to be more gender nonconforming as children and adults, and to prefer a receptive ASR as adults. In fact, a preference for a receptive ASR might reasonably be considered another aspect of gender nonconformity, perhaps directly influenced by androgen action on the developing brain or indirectly influenced via sociocultural expectations of gender nonconforming men. For exam-

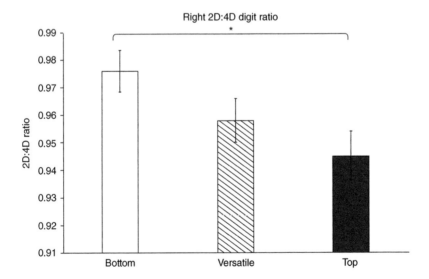

Figure 3. Differences in right-hand digit ratios in gay men based on their anal sex role (ASR) preference. ASR bottoms have more female-typical (higher) digit ratios than ASR tops ($P < 0.05$; $d = 0.62$). Versatiles displayed intermediate digit ratios, which were not significantly different from either of the other two groups. (Figure reprinted from Swift-Gallant et al. 2021 under a Creative Commons Attribution 4.0 International License.)

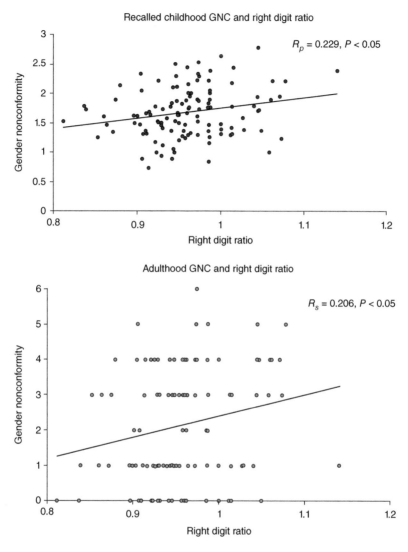

Figure 4. Right-hand digit ratios correlate with gender nonconformity scores. A more female-typical digit ratio is associated with higher gender nonconformity, in both childhood and adulthood, based on self-reports from gay men. (Figure reprinted from Swift-Gallant et al. 2021 under a Creative Commons Attribution 4.0 International License.)

ple, beliefs about masculinity may influence the degree to which gay men differentiate between receptive and insertive ASR as gendered behaviors (Ravenhill and de Visser 2018).

ISOLATED GnRH DEFICIENCY (IGD)

A rare endocrine condition offers a chance to partially dissociate the indirect influence of prenatal and neonatal gonadal steroids on psychol-

ogy via their effects on external appearance, and hence gender socialization, from the more direct influence via their effects on patterns of gene expression in the developing brain. In IGD, the GnRH neurons in the hypothalamus are either absent or nonfunctional (Laitinen et al. 2011). The absence of GnRH production has little effect on the early events of sexual differentiation because production of fetal gonadal steroids is initially driven by secretion of human chorionic gonadotropin (hCG) from the placenta (Braunstein and Hershman 1976). Because hCG and luteinizing hormone (LH) share a receptor, hCG binding to the shared hCG-LH receptor leads to the production of androgens and estrogens, and concomitantly, sex-typical sexual differentiation and development in both males and females. Thus, males with IGD are exposed to male-typical levels of androgens during early fetal development, but reduced levels in later fetal development and do not experience the postnatal "mini-puberty" from birth to approximately 3–6 months of age (Fig. 5, top), where testosterone levels approximate those in adulthood. Females with IGD are likewise exposed to female-typical levels of estrogens in early fetal development, but reduced levels in later fetal

development and no mini-puberty of varying levels of estrogen in the first 6 months of life (Fig. 5, bottom).

Investigation of people with IGD has found that the majority of both men and women are heterosexual. However, women with IGD were more likely to report bisexual orientation and were less androphilic than controls (Shirazi et al. 2021a), suggesting that perinatal estrogen action after the first trimester of gestation may normally potentiate androphilia.

Men with IGD report similar levels of gynephilia as control men but differ from control males in a different respect: they report greater childhood gender nonconformity than control males (Shirazi et al. 2021b). This effect is seen in both a population of men with IGD identified

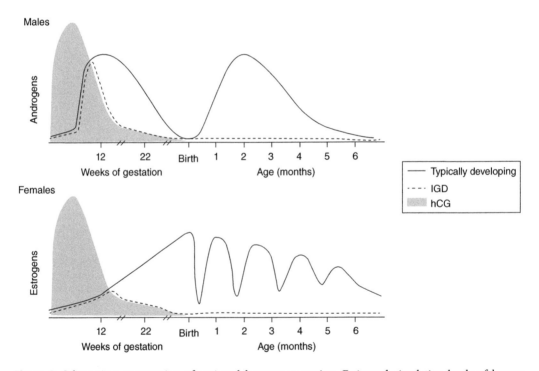

Figure 5. Schematic representation of perinatal hormone secretion. Estimated circulating levels of human chorionic gonadotropin (hCG; gray shading) and gonadal sex steroid hormone production in males (*top*) and females (*bottom*). In males, androgen production typically begins about the eighth week of gestation with the differentiation of the bipotential gonad (Siiteri and Wilson 1974; Negri-Cesi et al. 2004), persisting until the 24th week (Forest et al. 1973). In females, estrogen levels typically increase across gestation, peaking perinatally and fluctuating through mini-puberty, then ovarian activity ceases until puberty. In people with isolated gonadotropin-releasing hormone (GnRH) deficiency (IGD) (dotted lines), gonadal hormone production declines as hCG levels wane. (Figure from Shirazi et al. 2021a, adapted from Lanciotti et al. 2018, reprinted under the terms of the Creative Commons Attribution License (CC BY).)

from clinical diagnosis, and a sample of men recruited via the internet from IGD support groups who self-reported that they had been diagnosed with IGD (Fig. 6).

These data from people with IGD implicate the hormonal milieu from the period spanning approximately the second trimester of gestation through mini-puberty in solidifying sex-typical behaviors in both sexes. Importantly, as IGD is associated with sex-typical sexual development and usually not identified until young adulthood because of a delay in puberty, people with IGD are raised by parents and family who are usually unaware of their condition across infancy and childhood. Thus, differences between those with and without IGD are unlikely to be explained by effects of the condition on the social milieu, which might otherwise mediate the differences. Having minimized the chance that social influences are mediating these hormonal effects, we feel emboldened to assert that the perinatal hormones are acting directly upon the body and/or brain to affect these behaviors.

OVERALL CONCLUSIONS

Given the many social and cultural influences on every human behavior, including sexual behaviors, it may come as a surprise to think that perinatal hormones would have any effect. Yet the findings we reviewed above indicate that hormones do have an influence after all. It is difficult to explain how lesbians would come to have, on average, more masculine digit ratios than straight women any other way. Likewise, it is hard to discern why men with IGD would recall greater gender nonconformity than controls, or why women with IGD would be less androphilic, if their perinatal hormonal conditions were totally irrelevant to these behaviors. It seems that, in regard to human sexual and gendered behaviors, nature has a say after all, to borrow Fausto-Sterling's phrase (Fausto-Sterling 2000).

As we noted above, these data suggest that variation in perinatal hormone exposure promotes diversity in human psychology and be-

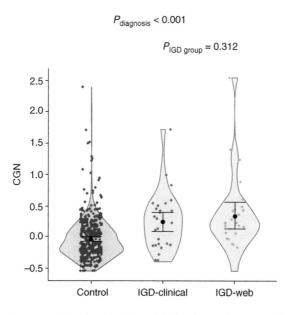

Figure 6. Childhood gender nonconformity (CGN) and isolated gonadotropin-releasing hormone (GnRH) deficiency (IGD). While men with IGD report similar levels of heterosexuality as control men, they do report greater gender nonconformity as children than control men. The differences seen in two samples of men with IGD, one a clinic-based group with diagnoses confirmed by physicians, and the other a web-based group of men self-identifying as having IGD, displayed the same pattern of results. (Figure reprinted from Shirazi et al. 2021b with permission from the authors.)

havior, including diversity in how fully people conform to social expectations of gendered behavior. Our perspective is that nature promotes an astonishing degree of diversity, and variation in the perinatal environment is simply one source of that diversity.

To avoid our being misunderstood, it might be helpful to explicate what these results do not mean. First and foremost, we are not suggesting that these findings offer any information about any particular individual. Knowing a particular woman's digit ratio offers little clue to her particular sexual orientation. Indeed, no matter what her digit ratio, she is almost certainly straight, since over 90% of women are. Nor does a gay man's digit ratio allow you to predict with any degree of certainty his preferred ASR. Meeting a man who reports that he has IGD does not let you predict with any certainty that he was unenthusiastic about playing sports as a boy. Likewise, knowing that a particular woman has IGD offers no reliable prediction about how sexually attracted she is to men or to women.

These caveats might seem apparent to anyone who has examined the degree of variance in our data, and they certainly are obvious to those of us who gathered them. But because they regard human behavior, perhaps particularly intimate and personal human behaviors, sometimes others make the logical leap that we are over-interpreting our data so far as to explain, for example, why one woman is straight while another is not. Some of the associations reviewed above, particularly those concerning digit ratios, rely on moderate-to-large samples of people, which might suggest that these hormonal effects are subtle, although the noisiness of digit ratio as a marker for prenatal androgen signaling is also likely to contribute to modest statistical associations. Yet even where associations suggest stronger effects of perinatal steroids, such as those from studies of people with CAH and IGD, the finding that nature has a say does not disprove the important role of social interactions, cultural influences, experience, or a multitude of other factors that affect behavior. We feel confident that society also has a say in explaining variation in these psychological and behavioral traits. Indeed, one of the most important ways that gonadal steroids may influence

human behavior is to affect social interactions—which individuals a person pays attention to and learns from, as well as behavioral patterns that influence how a person is treated by others, for example (e.g., Hines et al. 2016). Put another way, human behavior is overdetermined, sensitive to myriad, and highly variable influences.

Another important limitation of these data is that they say nothing about the site of action of steroids on these behaviors. One might assume that if hormones affect behavior, then they must do so by acting directly upon the brain, and that social factors were irrelevant to their action. In laboratory animals, sex hormones indeed act directly on the developing brain, regulating patterns of gene expression in ways that lead to sex differences in behavior (Xu et al. 2012). We also know that there are sex differences in the expression of thousands of genes in human brains (Oliva et al. 2020). One might then infer that sex hormones also engender behavioral sex differences in humans by modulating patterns of gene expression in the brain, as they largely do in other mammals. However, none of the findings we have reviewed address the site of action, and studies in laboratory animals do support a role for sex hormone action outside of the brain (for review, see Monks and Swift-Gallant 2018). To address site of action in humans, we would have to find some means of selectively measuring hormone exposure or response in the brain versus other parts of the body, and we know of no such means. If perinatal hormones affect other parts of the body, and we know they do, then these influences on behavior could be mediated by those sites of action. For example, if prenatal androgens affect facial characteristics in addition to digit ratios, then perhaps girls who vary in these facial features may elicit different social stimuli that nudge them into either a gay or straight orientation during development. Similarly, maybe among boys following a developmental trajectory toward becoming gay, some differ in their muscle development, or degree of proprioception, that leads them to be more gender nonconforming than others.

We find these questions interesting and of course hope that others will as well. One characteristic that distinguishes humans from other animals is the degree of effort we are willing to expend to gain a greater understanding of ourselves as individuals and as a species. We note the special significance sex and sexuality: There could scarcely be a facet of social life more central than sexuality, and there is no single variable more important in explaining individual variation than biological sex. Understanding sexuality and sex differences is thus fundamental to clarifying social processes and individual variations in them. Our modest contribution to this enterprise includes these findings suggesting that perinatal hormones affect human sexual orientation, sexual behavior preferences, and childhood gender conformity.

REFERENCES

Atkinson BM, Smulders TV, Wallenberg JC. 2017. An endocrine basis for tomboy identity: the second-to-fourth digit ratio (2D:4D) in "tomboys." *Psychoneuroendocrinology* 79: 9–12. doi:10.1016/j.psyneuen.2017.01.032

Barrett ES, Parlett LE, Swan SH. 2015. Stability of proposed biomarkers of prenatal androgen exposure over the menstrual cycle. *J Dev Orig Health Dis* 6: 149–157. doi:10.1017/S2040174414000646

Baxter A, Wood EK, Witczak LR, Bales KL, Higley JD. 2020. Sexual dimorphism in titi monkeys' digit (2D:4D) ratio is associated with maternal urinary sex hormones during pregnancy. *Dev Psychobiol* 62: 979–991. doi:10.1002/dev.21899

Berenbaum SA, Bryk KK, Nowak N, Quigley CA, Moffat S. 2009. Fingers as a marker of prenatal androgen exposure. *Endocrinology* 150: 5119–5124. doi:10.1210/en.2009-0774

Braunstein GD, Hershman JM. 1976. Comparison of serum pituitary thyrotropin and chorionic gonadotropin concentrations throughout pregnancy. *J Clin Endocrinol Metab* 42: 1123–1126. doi:10.1210/jcem-42-6-1123

Brown WM, Finn CJ, Cooke BM, Breedlove SM. 2002a. Differences in finger length ratios between self-identified "butch" and "femme" lesbians. *Arch Sex Behav* 31: 123–127. doi:10.1023/A:1014091420590

Brown WM, Hines M, Fane BA, Breedlove SM. 2002b. Masculinized finger length patterns in human males and females with congenital adrenal hyperplasia. *Horm Behav* 42: 380–386. doi:10.1006/hbeh.2002.1830

Buck JJ, Williams RM, Hughes IA, Acerini CL. 2003. In-utero androgen exposure and 2nd to 4th digit length ratio-comparisons between healthy controls and females with classical congenital adrenal hyperplasia. *Hum Reprod* 18: 976–979. doi:10.1093/humrep/deg198

Chang S, Skakkebaek A, Trolle C, Bojesen A, Hertz JM, Cohen A, Hougaard DM, Wallentin M, Pedersen AD, Østergaard JR, et al. 2015. Anthropometry in Klinefelter syndrome—multifactorial influences due to CAG length, testosterone treatment and possibly intrauterine hypogo-

nadism. *J Clin Endocrinol Metab* 100: E508–E517. doi:10.1210/jc.2014-2834

Fausto-Sterling A. 2000. *Sexing the body*. Basic Books, New York.

Forest MG, Cathiard AM, Bertrand JA. 1973. Evidence of testicular activity in early infancy. *J Clin Endocrinol Metab* 37: 148–151. doi:10.1210/jcem-37-1-148

Galis F, Ten Broek CM, Van Dongen S, Wijnaendts LC. 2010. Sexual dimorphism in the prenatal digit ratio (2D:4D). *Archives Sexual Behavior* 39: 57–62. doi:10.1007/s10508-009-9485-7

Grimbos T, Dawood K, Burriss RP, Zucker KJ, Puts DA. 2010. Sexual orientation and the second to fourth finger length ratio: a meta-analysis in men and women. *Behav Neurosci* 124: 278–287. doi:10.1037/a0018764

Hall LS, Love CT. 2003. Finger-length ratios in female monozygotic twins discordant for sexual orientation. *Arch Sex Behav* 32: 23–28. doi:10.1023/A:1021837211630

Hines M. 2011. Prenatal endocrine influences on sexual orientation and on sexually differentiated childhood behavior. *Front Neuroendocrinol* 32: 170–182. doi:10.1016/j.yfrne.2011.02.006

Hines M, Pasterski V, Spencer D, Neufeld S, Patalay P, Hindmarsh PC, Hughes IA, Acerini CL. 2016. Prenatal androgen exposure alters girls' responses to information indicating gender-appropriate behaviour. *Philos Trans R Soc Lond B Biol Sci* 371: 20150125. doi:10.1098/rstb.2015.0125

Hiraishi K, Sasaki S, Shikishima C, Ando J. 2012. The second to fourth digit ratio (2D:4D) in a Japanese twin sample: heritability, prenatal hormone transfer, and association with sexual orientation. *Arch Sex Behav* 41: 711–724. doi:10.1007/s10508-011-9889-z

Hönekopp J, Watson S. 2010. Meta-analysis of digit ratio 2D:4D shows greater sex difference in the right hand. *Am J Hum Biol* 22: 619–630. doi:10.1002/ajhb.21054

Hughes IA, Deeb A. 2006. Androgen resistance. *Best Pract Res Clin Endocrinol Metab* 20: 577–598. doi:10.1016/j.beem.2006.11.003

Kraemer B, Noll T, Delsignore A, Milos G, Schnyder U, Hepp U. 2006. Finger length ratio (2D:4D) and dimensions of sexual orientation. *Neuropsychobiology* 53: 210–214. doi:10.1159/000094730

Laitinen EM, Vaaralahti K, Tommiska J, Eklund E, Tervaniemi M, Valanne L, Raivio T. 2011. Incidence, phenotypic features and molecular genetics of Kallmann syndrome in Finland. *Orphanet J Rare Dis* 6: 41. doi:10.1186/1750-1172-6-41

Lanciotti L, Cofini M, Leonardi A, Penta L, Esposito S. 2018. Up-to-date review about minipuberty and overview on hypothalamic-pituitary-gonadal axis activation in fetal and neonatal life. *Front Endocrinol (Lausanne)* 9: 410. doi:10.3389/fendo.2018.00410

Leslie M. 2019. The mismeasure of hands? *Science* 364: 923–925. doi:10.1126/science.364.6444.923

Lippa RA. 2003. Are 2D:4D finger-length ratios related to sexual orientation? Yes for men, no for women. *J Pers Soc Psychol* 85: 179–188. doi:10.1037/0022-3514.85.1.179

Malas MA, Dogan S, Evcil EH, Desdicioglu K. 2006. Fetal development of the hand, digits and digit ratio (2D:4D).

Early Hum Dev **82:** 469–475. doi:10.1016/j.earlhumdev.2005.12.002

Manning JT, Scutt D, Wilson J, Lewis-Jones DI. 1998. The ratio of 2nd to 4th digit length: a predictor of sperm numbers and concentrations of testosterone, luteinizing hormone and oestrogen. *Hum Reprod* **13:** 3000–3004. doi:10.1093/humrep/13.11.3000

Manning JT, Kilduff LP, Trivers R. 2013. Digit ratio (2D:4D) in Klinefelter's syndrome. *Andrology* **1:** 94–99. doi:10.1111/j.2047-2927.2012.00013.x

McCarthy MM, Arnold AP, Ball GF, Blaustein JD, De Vries GJ. 2012. Sex differences in the brain: the not so inconvenient truth. *J Neurosci* **32:** 2241–2247. doi:10.1523/JNEUROSCI.5372-11.2012

McFadden D, Pasanen EG. 1998. Comparison of the auditory systems of heterosexuals and homosexuals: click-evoked otoacoustic emissions. *Proc Natl Acad Sci* **95:** 2709–2713. doi:10.1073/pnas.95.5.2709

McFadden D, Shubel E. 2002. Relative lengths of fingers and toes in human males and females. *Horm Behav* **42:** 492–500. doi:10.1006/hbeh.2002.1833

McFadden D, Loehlin JC, Breedlove SM, Lippa RA, Manning JT, Rahman Q. 2005. A reanalysis of five studies on sexual orientation and the relative length of the 2nd and 4th fingers (the 2D:4D ratio). *Arch Sex Behav* **34:** 341–356. doi:10.1007/s10508-005-3123-9

McFadden D, Pasanen EG, Valero MD, Roberts EK, Lee TM. 2009. Effect of prenatal androgens on click-evoked otoacoustic emissions in male and female sheep (*Ovis aries*). *Horm Behav* **55:** 98–105. doi:10.1016/j.yhbeh.2008.08.013

Monks DA, Swift-Gallant A. 2018. Non-neural androgen receptors affect sexual differentiation of brain and behaviour. *J Neuroendocrinol* **30:** e12493. doi:10.1111/jne.12493

Moskowitz DA, Hart TA. 2011. The influence of physical body traits and masculinity on anal sex roles in gay and bisexual men. *Arch Sex Behav* **40:** 835–841. doi:10.1007/s10508-011-9754-0

Nave G, Koppin CM, Manfredi D, Richards G, Watson SJ, Geffner ME, Yong JE, Kim R, Ross HM, Serrano-Gonzalez M, et al. 2021. No evidence for a difference in 2D:4D ratio between youth with elevated prenatal androgen exposure due to congenital adrenal hyperplasia and controls. *Horm Behav* **128:** 104908. doi:10.1016/j.yhbeh.2020.104908

Negri-Cesi P, Colciago A, Celotti F, Motta M. 2004. Sexual differentiation of the brain: role of testosterone and its active metabolites. *J Endocrinol Invest* **27:** 120–127.

O'Kelly F, DeCotiis K, Zu'bi F, Farhat WA, Koyle MA. 2020. Increased hand digit length ratio (2D:4D) is associated with increased severity of hypospadias in pre-pubertal boys. *Pediatr Surg Int* **36:** 247–253. doi:10.1007/s00383-019-04600-3

Ökten A, Kalyoncu M, Yariş N. 2002. The ratio of second- and fourth-digit lengths and congenital adrenal hyperplasia due to 21-hydroxylase deficiency. *Early Hum Dev* **70:** 47–54. doi:10.1016/S0378-3782(02)00073-7

Oliva M, Muñoz-Aguirre M, Kim-Hellmuth S, Wucher V, Gewirtz ADH, Cotter DJ, Parsana P, Kasela S, Balliu B, Viñuela A, et al. 2020. The impact of sex on gene expression across human tissues. *Science* **369:** eaba3066. doi:10.1126/science.aba3066

Oswiecimska JM, Ksiazek A, Sygulla K, Pys-Spychala M, Roczniak GR, Roczniak W, Stojewska M, Ziora K. 2012. Androgens concentrations and second-to fourth-digit ratio (2D:4D) in girls with congenital adrenal hyperplasia (21-hydroxylase deficiency). *Neuro Endocrinol Lett* **33:** 787–791.

Phoenix CH, Goy RW, Gerall AA, Young WC. 1959. Organizing action of prenatally administered testosterone propionate on the tissues mediating mating behavior in the female Guinea pig. *Endocrinology* **65:** 369–382. doi:10.1210/endo-65-3-369

Putz DA, Gaulin SJ, Sporter RJ, McBurney DH. 2004. Sex hormones and finger length: what does 2D:4D indicate? *Evol Hum Behav* **25:** 182–199. doi:10.1016/j.evolhumbehav.2004.03.005

Rahman Q. 2005. Fluctuating asymmetry, second to fourth finger length ratios and human sexual orientation. *Psychoneuroendocrinology* **30:** 382–391. doi:10.1016/j.psyneuen.2004.10.006

Rahman Q, Wilson GD. 2003. Sexual orientation and the 2nd to 4th finger length ratio: evidence for organising effects of sex hormones or developmental instability? *Psychoneuroendocrinology* **28:** 288–303. doi:10.1016/S0306-4530(02)00022-7

Ravenhill JP, de Visser RO. 2018. "It takes a man to put me on the bottom": gay men's experiences of masculinity and anal intercourse. *J Sex Res* **55:** 1033–1047. doi:10.1080/00224499.2017.1403547

Rivas MP, Moreira LM, Santo LD, Marques AC, El-Hani CN, Toralles MB. 2014. New studies of second and fourth digit ratio as a morphogenetic trait in subjects with congenital adrenal hyperplasia. *Am J Hum Biol* **26:** 559–561. doi:10.1002/ajhb.22545

Robinson SJ, Manning JT. 2000. The ratio of 2nd to 4th digit length and male homosexuality. *Evol Hum Behav* **21:** 333–345. doi:10.1016/S1090-5138(00)00052-0

Shirazi TN, Self H, Dawood K, Welling LLM, Cárdenas R, Rosenfield KA, Bailey JM, Balasubramanian R, Delaney A, Breedlove SM, et al. 2021a. Evidence that perinatal ovarian hormones promote women's sexual attraction to men. *Psychoneuroendocrinol* **134:** 105431. doi:10.1016/j.psyneuen.2021.105431

Shirazi TN, Self H, Dawood K, Welling LLM, Cárdenas R, Rosenfield KA, Bailey JM, Balasubramanian R, Delaney A, Crowley W, et al. 2021b. Perinatal gonadal hormones, childhood gender nonconformity and sexual orientation: evidence from isolated GnRH deficiency. *Psychol Sci* (in press).

Siiteri PK, Wilson JD. 1974. Testosterone formation and metabolism during male sexual differentiation in the human embryo. *J Clin Endocrinol Metab* **38:** 113–125. doi:10.1210/jcem-38-1-113

Skorska MN, Bogaert AF. 2017. Prenatal androgens in men's sexual orientation: evidence for a more nuanced role? *Arch Sex Behav* **46:** 1621–1624. doi:10.1007/s10508-017-1000-y

Spencer D, Pasterski V, Neufeld SAS, Glover V, O'Connor TG, Hindmarsh PC, Hughes IA, Acerini CL, Hines M. 2021. Prenatal androgen exposure and children's gender-

Cite this article as *Cold Spring Harb Perspect Biol* doi: 10.1101/cshperspect.a039123

typed behavior and toy and playmate preferences. *Horm Behav* **127**: 104889. doi:10.1016/j.yhbeh.2020.104889

Swift-Gallant A, Coome LA, Monks DA, VanderLaan DP. 2018. Gender nonconformity and birth order in relation to anal sex role among gay men. *Arch Sex Behav* **47**: 1041–1052. doi:10.1007/s10508-017-0980-y

Swift-Gallant A, Di Rita V, Major CA, Breedlove CJ, Jordan CL, Breedlove SM. 2021. Differences in digit ratios between gay men who prefer receptive versus insertive sex roles indicate a role for prenatal androgen. *Sci Rep* **11**: 8102. doi:10.1038/s41598-021-87338-0

Tortorice JL. 2002. "Written on the body: butch/femme lesbian gender identity and biological correlates." Thesis/dissertation, Rutgers University, New Brunswick, NJ.

van Hemmen J, Cohen-Kettenis PT, Steensma TD, Veltman DJ, Bakker J. 2017. Do sex differences in CEOAEs and 2D:4D ratios reflect androgen exposure? A study in women with complete androgen insensitivity syndrome. *Biol Sex Differ* **8**: 11. doi:10.1186/s13293-017-0132-z

Wallien MS, Zucker KJ, Steensma TD, Cohen-Kettenis PT. 2008. 2D:4D finger-length ratios in children and adults

with gender identity disorder. *Horm Behav* **54**: 450–454. doi:10.1016/j.yhbeh.2008.05.002

Watts TM, Holmes L, Raines J, Orbell S, Rieger G. 2018. Finger length ratios of identical twins with discordant sexual orientations. *Arch Sex Behav* **47**: 2435–2444. doi:10.1007/s10508-018-1262-z

Williams TJ, Pepitone ME, Christensen SE, Cooke BM, Huberman AD, Breedlove NJ, Breedlove TJ, Jordan CL, Breedlove SM. 2000. Finger-length ratios and sexual orientation. *Nature* **404**: 455–456. doi:10.1038/35006555

Xu X, Coats JK, Yang CF, Wang A, Ahmed OM, Alvarado M, Izumi T, Shah NM. 2012. Modular genetic control of sexually dimorphic behaviors. *Cell* **148**: 596–607. doi:10.1016/j.cell.2011.12.018

Zheng Z, Cohn MJ. 2011. Developmental basis of sexually dimorphic digit ratios. *Proc Natl Acad Sci* **108**: 16289–16294. doi:10.1073/pnas.1108312108

Zheng L, Hart TA, Zheng Y. 2012. The relationship between intercourse preference positions and personality traits among gay men in China. *Arch Sex Behav* **41**: 683–689. doi:10.1007/s10508-011-9819-0

Integrating Sex Chromosome and Endocrine Theories to Improve Teaching of Sexual Differentiation

Arthur P. Arnold

Department of Integrative Biology & Physiology, and Laboratory of Neuroendocrinology of the Brain Research Institute, University of California, Los Angeles, California 90095-7239, USA

Correspondence: arnold@ucla.edu

Major sex differences in mammalian tissues are functionally tied to reproduction and evolved as adaptations to meet different reproductive needs of females and males. They were thus directly controlled by gonadal hormones. Factors encoded on the sex chromosomes also cause many sex differences in diverse tissues because they are present in different doses in XX and XY cells. The sex chromosome effects likely evolved not because of demands of reproduction, but as side effects of genomic forces that adaptively reduced sexual inequality. Sex-specific effects of particular factors, including gonadal hormones, therefore, are not necessarily explained as adaptations for reproduction, but also as potential factors offsetting, rather than producing, sex differences. The incorporation of these concepts would improve future teaching about sexual differentiation.

For more than 100 years, gonadal hormones have been implicated as the cause of sexually dimorphic development of various tissues (Cooke et al. 1998; Arnold 2002). More recently, X and Y genes have joined the list of factors that contribute to sexual differentiation. Here, we pursue the idea that consideration of evolution of the sex chromosomes considerably enriches our understanding of natural sex differences in physiology and disease. Many unequal effects of X and Y genes did not evolve primarily as adaptations promoting female or male forms of reproduction, and therefore may be outside of the conceptual framework of many scientists studying sexual differentiation caused by gonadal hormones. New students of sexual differentiation may benefit from an appreciation not only of the evolution of endocrine control of sexual dimorphism, but also the evolution of sexual imbalance and balance in the sex chromosomes. The combined appreciation of divergent classes of sex-biasing factors will be required for understanding their interaction in complex control of sex differences in physiology and disease.

Experiments of the twentieth century established a two-step model of sexual development in mammals (Fig. 1). The first step, called "sex determination," is genetic (Lillie 1939; Achermann and Jameson 2017). In XY embryos, the Y-linked testis-determining gene, *Sry*, is expressed by midgestation within the undifferentiated gonadal ridge where it causes the development of testes.

Twentieth century hormonal theory

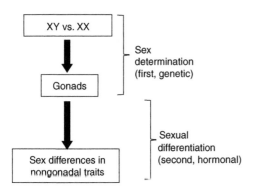

Figure 1. Two-step hormonal theory of sexual differentiation. The dominant theory of sexual differentiation developed in the twentieth century proposed a two-step process (Lillie 1939). In XY embryos, cell-autonomous expression of Y-linked *Sry* in the undifferentiated gonad caused differentiation of testes, whereas autosomal genes initiated differentiation of ovaries in XX embryos. This first step was called "sex determination." Then, a second step occurred, called "sexual differentiation," in which gonadal hormones caused sex differences by acting on nongonadal tissues.

The testes make sperm and secrete predominantly testosterone. In XX embryos that lack *Sry*, autosomal and X chromosome genes cause development of ovaries (Capel 2017), which makes eggs and secrete predominantly estrogens and progesterone. Because of the centrality of gonad and gamete type in the definition of sex of any sexual species (males make sperm, females make eggs), these genetic events were seen to determine the biological sex of the individual. Following differentiation of the gonads, the second step occurs, called "sexual differentiation," in which different secretions of ovaries or testes cause unequal sex-specific development of phenotypes in nongonadal tissues. These ideas were proposed by pioneers such as Lillie (1939) and critically tested by the middle of the twentieth century by Jost (1970) and others. This conceptual framework was refined by experiments such as those of Phoenix et al. (1959), who distinguished between permanent or differentiating actions of testosterone from the testes ("organizational" effects, especially at early stages of or-

gan development), and the acute, reversible actions of any of the gonadal hormones throughout life ("activational" effects) (Arnold 2009; Bakker and Brock 2010). Of course, the hormonal control of sexual differentiation involves hormonal regulation of many downstream molecular and cellular events, which are mediated by genes throughout the genome. The two-step model of sexual development has been useful to clinicians, who discriminate clinical syndromes resulting from disruption of the genetic pathways regulating gonadal development from those disrupting the hormonal actions regulating sexual phenotype in nongonadal tissues (Achermann and Jameson 2017). The two-step model is ingrained in the educational framework, and is taught in many undergraduate and graduate institutions and medical schools.

In the last 20 years, however, the two-step model has been undermined as significant new evidence has emerged that phenotypic sex differences in virtually all tissues of the body are caused in part by inherently unequal effects of X and Y genes expressed in XX and XY cells (Arnold 2019). These cell-autonomous "sex chromosome effects" disprove the idea that the inequality in effects of testicular and ovarian secretions are the only source of phenotypic sex differences outside of the gonad. Sex differences in gene expression begin as soon as the embryonic genome begins transcription, long before gonadal differentiation (Bermejo-Alvarez et al. 2011; Lowe et al. 2015). In the mouse heart, for example, the number of X chromosomes contributes to sex differences in expression of proteins, and hundreds of proteins show sex differences in expression prior to differentiation of the gonads (Shi et al. 2021).

EVIDENCE FOR CELL-AUTONOMOUS SEX CHROMOSOME EFFECTS CAUSING SEXUAL DIFFERENTIATION

Although studies of numerous species contribute to the idea that sex differences in tissues are caused in part by sex chromosome effects (Renfree and Short 1988; Arnold et al. 2013), by now most evidence comes from the study of mouse models with unusual sex chromosomes. The "four core genotypes" (FCG) model involves deletion of the testis-determining *Sry* gene from the Y chromosome, together with insertion of an *Sry* transgene onto chromosome 3 (De Vries et al. 2002; Burgoyne and Arnold 2016). These manipulations make the type of gonad independent of sex chromosome complement (XX vs. XY). Four genotypes are produced: XX and XY gonadal males with *Sry*, and XX and XY gonadal females lacking *Sry*. The model allows detection of hormonally controlled sex differences (phenotypic differences between mice with testes vs. ovaries), sex chromosome effects (phenotypic differences between XX and XY mice with the same type of gonad), and the interaction of the two factors. A second mouse model, the XY* model, compares mice with the same gonad type but with different numbers of X chromosomes while keeping the number of Y chromosomes constant (XO vs. XX gonadal females, XY vs. XXY gonadal males), to measure the influence of X chromosome dose on traits. The XY* model also gives information about the effect of the Y chromosome, comparing mice with and without the Y chromosome but with the same number of X chromosomes (XO vs. XY, XX vs. XXY) (Burgoyne and Arnold 2016; Arnold 2020).

Three recent studies exemplify demonstrations of sex chromosome effects. Corre et al. (2016) performed structural magnetic resonance imaging (MRI) on the brains of FCG mice, to discover brain regions showing sex differences in regional brain volumes that can be attributed to sex hormone or sex chromosome effects, or both. In a segmented atlas producing measurements of 62 brain regions, 30 regions showed sex differences. Of these, 16 depended on the type of gonad, 11 showed effects of sex chromosome complement, and only three showed both types of effects. The size of sex differences caused by each type of variable was comparable and persisted in mice that were gonadectomized before puberty (Vousden et al. 2018). This study challenges the long-held view that the large majority of sex differences in the brain are caused by gonadal hormone effects (McCarthy and Arnold 2011).

A second study shows sex chromosome effects on antinociception in tests of pain behavior

(Taylor et al. 2022). The level of nociception was measured by the latency of a mouse to withdraw its tail from hot water. The effectiveness of κ opioid analgesics was tested in wild-type, FCG, and XY* mice (Fig. 2). Females show reduced antinociception caused by κ antagonists, relative to males. In FCG mice, XX mice showed reduced antinociception compared to XY mice, but the type of gonad had no effect. The sex chromosome effect was confirmed both in gonad-intact FCG mice and FCG mice gonadectomized in adulthood. The sex chromosome effect was also confirmed in the completely independent XY* model, which lacks any transgene, with the result that mice with two X chromosomes (irrespective of gonad type) showed reduced antinociception relative to mice with one X chromosome. These studies rationalize the search for X genes causing sex differences in κ opioid antinociception because of constitutive differences in expression in XX and XY cells. As in other studies of FCG and XY* mice, the sex chromosome effects were discovered under conditions in which the effects are unlikely to be explained by hormonal differences between sex chromosome groups. XX and XY FCG groups with the same type of gonad have comparable anogenital distance and adult hormone levels, suggesting that both prenatal and adult hormone levels are comparable in XX and XY mice with the same type of gonad (Burgoyne and Arnold 2016). The sex chromosome effect is found in mice lacking gonadal hormones as adults, after gonadectomy, ruling out differences in the levels of gonadal hormones in XX and XY groups at the time of testing. The lack of a difference between groups born with ovaries versus testes argues against a hormonal effect on the phenotype under the conditions of testing. And the sex chromosome effects are confirmed in mice with ovaries and with testes, suggesting that the sex chromosome effects are robust in the face of major differences in background levels of gonadal hormones.

The third example shows an effect of sex chromosome complement in a model of brain disease (Davis et al. 2020). One mouse model of Alzheimer's disease (AD) involves introduction of the gene encoding human amyloid precursor protein (hAPP) into the mouse genome,

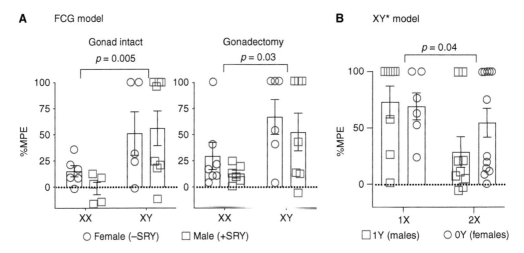

Figure 2. Sex chromosomes contribute to sex differences in κ opioid antinociception. The response to pain was measured by the latency with which mice withdraw their tail from hot water. The graphs show maximum possible effect (MPE) of a U50,488H, a κ opioid receptor agonist that is antinociceptive and increases the latency. In wild-type mice, females show less antinociceptive response to the drug than males (not shown). (A) In four core genotypes (FCG) mice, XX mice show less antinociception than XY mice, whether they are gonad-intact or gonadectomized in adulthood. The XX versus. XY difference is seen in mice with ovaries (females, lacking *Sry*) or with testes (males, with *Sry*). (B) In the XY* mouse model, gonadal female and male mice with two X chromosomes (XX, XXY, respectively) show less antinociception than mice with one X chromosome (XO, XY). The results implicate gene(s) on the X chromosome as factors causing the sex difference. (Figure from Taylor et al. 2020; reprinted, with permission, from John Wiley and Sons © 2020.)

which leads to premature death and some cognitive deficits reminiscent of AD. Male hAPP mice die earlier than females, whether they are gonad-intact or gonadectomized. In FCG groups, XY mice die earlier than XX mice, a main effect of sex chromosomes. But the type of gonad also influences longevity, producing earlier death in gonadal females than gonadal males in XY but not XX mice. In the XY* model, mice with one X chromosome die earlier than those with two X chromosomes (main effect of X chromosome number), but again the presence of testes is protective in mice with one X chromosome (but not in those with two X chromosomes). The authors provided evidence that the X-linked gene *Kdm6a* contributes to the sex chromosome effect, protecting XX females because two doses of *Kdm6a* are present, relative to males when only one dose is present. The protective effects of greater *Kdm6a* dose are demonstrated in vitro, in studies of β-amyloid toxicity, and in vivo, measuring protective effects of *Kdm6a* dose on learning in a Morris water maze in hAPP mice. Thus, the constitutive sex difference in a dose of an X gene, dependent on the number of X chromosomes, contributes to sex differences in a mouse model of AD.

Including these examples, sex chromosome effects have by now been reported in mice to cause sex differences in brain morphology and function (Carruth et al. 2002; Chen et al. 2009; Cox et al. 2014; Corre et al. 2016; Vousden et al. 2018), in behavior (Gioiosa et al. 2008; Cox and Rissman 2011; Bonthuis et al. 2012; Kuljis et al. 2013; Cox et al. 2014, 2015; Chen et al. 2015; Taylor et al. 2022), and in mouse models of autoimmune disease (Palaszynski et al. 2005; Smith-Bouvier et al. 2008; Case et al. 2013; Du et al. 2014; Itoh et al. 2019), neurodegeneration (Du et al. 2014), metabolism (Chen et al. 2012, 2013; Link et al. 2015, 2017, 2020), numerous cardiovascular diseases (systemic and pulmonary hypertension, stroke, ischemia/reperfusion injury, aneurysms, atherosclerosis) (Ji et al. 2010; Li et al. 2014; McCullough et al. 2016; Alsiraj et al. 2017, 2018, 2019; Arnold et al. 2017), cancer

(Kaneko and Li 2018), pain (Gioiosa et al. 2008; Taylor et al. 2022), addiction (Quinn et al. 2007; Barker et al. 2010), and neural tube closure defects (Chen et al. 2008). This list is not comprehensive but is limited by the number of experiments that have sought to find sex chromosome effects (Arnold 2020).

DIVERGENCE OF GENETIC AND HORMONAL CONCEPTUAL FRAMEWORKS AND RESEARCH COMMUNITIES

As scientists built up confidence and embraced the two-step model (genetic sex determination followed by hormonal sexual differentiation), it affected many decisions in the conduct of science and design of experiments, and in education, publishing, and support of research. The scientists studying gonadal (genetic) and nongonadal (hormonal) sexual differentiation became relatively isolated from each other. Investigators of gonadal sexual differentiation viewed themselves as molecular developmental biologists, whereas those studying nongonadal sexual differentiation viewed themselves as endocrinologists. Each group trained their students in relative isolation of the other. The group focused on genetics and molecular biology of gonadal differentiation often were interested in the sex chromosomes, because gonadal development stemmed so directly from expression of Y or X genes that were constitutively out of balance in XX and XY cells. In scientific societies having to do with nongonadal tissues (e.g., Society for Neuroscience), sexual differentiation was a subtopic under endocrinology. For example, students of neuroendocrinology learned principles of endocrinology but had little exposure to the biology of sex chromosomes. The endocrine focus of neuroendocrinology significantly influenced the choice of experiments that were proposed and performed and the research grants that were awarded.

The two-step theory, and the proposed centrality of the gonad for sexual differentiation, influenced the interpretation of experimental results. Some events that occur in typical females but not in typical males were not viewed as part of sexual differentiation if they were not up-

stream or downstream of gonadal development, even though they were clearly sexually differentiated (Arnold et al. 2013; Arnold 2019). For example, the female-specific expression of *Xist* is a sexual dimorphism itself and has sex-biasing effects. Female mice require *Xist* to prevent hematologic cancer, but males do not (Yildirim et al. 2013), and disease mechanisms interact with *Xist* to have sex-biased effects (Delbridge et al. 2019; Yu et al. 2021). As a second example, men are color-blind more often than women because of genetic variations in X-linked opsin genes that have larger effects in XY cells, where only one allele is present, compared to XX cells where the effect is offset by the presence of a second allele (Midgeon 2007). Because the mechanism causing this sex difference was unrelated to gonadal development, color-blindness or other male-biased X-linked diseases (Fragile X Syndrome, hemophilia, etc.) were rarely considered in theoretical formulations of sexual differentiation and were not seen as exceptions to the two-step theory.

The hormonal conceptual framework also influenced thinking about how and why sex differences evolve. Early studies of sexually differentiated tissues were aimed at understanding development of explicitly reproductive functions. Lillie and Jost studied factors causing different development of internal and external genitals in the two sexes, tissues that were required for the individual to play a male or female reproductive role. The earliest studies of sexual differentiation of behavior focused on copulatory and courtship behaviors, essential for reproduction. It was reasonable to imagine that male copulatory behavior evolved sensitivity to testicular secretions because it was adaptive and promoted better male reproduction. Testosterone was considered a hormonal signal that coordinated testicular activity with the neural functions required for fertility of the male. Each sex difference was seen to have evolved because it served the different roles required for reproduction in males or females. For example, in species with territorial aggression performed by males, which was organized and activated by testicular androgens, it was easy to postulate that the sex differences evolved because testosterone promoted the behaviors that assured defense of resources (mate, food, shelter) required for successful development and survival of the male's offspring.

A potentially major issue was that sex steroid–induced sex differences also occurred in tissues (e.g., liver, adipose) or processes that were not directly related to reproduction. One resolution was to generalize the idea that tissue functions, regulated by gonadal hormones, evolved hormonal sensitivity because these supposedly "nonreproductive" tissues nevertheless had sex-specific roles in the reproductive lifetime of the individual. In other words, even nonreproductive tissues were adapted for better reproductive function. As just one example, in a recent review concerning sex differences in metabolism, estrogen effects on the liver were interpreted as an adaptation to the demands of the female's role in reproduction. "Mammals are characterized by a major change in reproductive strategies and it is conceivable that these changes subjected females to a significant evolutionary pressure that perfected the coupling between energy metabolism and reproduction" (Maggi and Della Torre 2018). This is certainly a plausible hypothesis. Although there can be no doubt that mechanisms of sexual differentiation of metabolism must have evolved based on selection pressures that require compatibility with female (or male) reproduction, competing narratives offer other explanations of evolution of sex differences that are unrelated to reproduction (see below). The dominant attribution of sex differences to effects of gonadal hormones might limit the range of evolutionary scenarios considered. Gonadal steroid hormones have so many diverse effects that it is not always easy to imagine why a specific example of sensitivity to gonadal hormones evolved as an adaptation to more successful reproduction. Moreover, as discussed further below, many or most sex-biasing effects of sex chromosome genes did not evolve to optimize female or male reproduction but as the result of forces of genomic evolution. The inherent imbalance of sex chromosome gene actions in all somatic cells was often a problem, not a solution, which favored evolution of other sex differences to offset them (De Vries 2004). For example, gonadal hormone effects may have evolved to offset maladaptive sex biases that occur either because of sex chromosome effects or diverse pleiotropic effects ("side effects") of individual gonadal hormones.

EVOLUTION OF SEX DIFFERENCES IN THE GENOME

Although numerous vertebrate species lack sex chromosomes, with gonad type being regulated by environmental factors such as temperature, we concentrate here on vertebrate species with heteromorphic sex chromosomes and genetic control of gonadal differentiation (so-called "genetic sex determination"). Sex chromosomes have evolved repeatedly and independently in plants and animals (Charlesworth 2003). Although numerous variations exist, many species, such as mammals and birds, have one large sex chromosome and one small sex chromosome. The homogametic sex has two copies of the large sex chromosome (XX females in mammals or ZZ males in birds), and the heterogametic sex has one copy of each (XY males in mammals or ZW females in birds). These chromosomes evolved from an ancient autosomal pair. On one of the chromosomes, mutations led to the emergence of a dominant gonad-determining factor (Charlesworth et al. 2005). During emergence of eutherian and marsupial (metatherian) mammals, the gene Sox3 sustained mutations leading to the evolution of Sry, which acquired the role of initiating development of testes (Sutton et al. 2011). The Sry-containing chromosome became the incipient Y chromosome, and its previous autosomal partner evolved into the X chromosome. The emergence of Sry triggered a chain of evolutionary events that resulted in major alterations in the genome, dramatically affecting both the Y and X chromosomes, but also the autosomes that were required to adjust to evolutionary changes in the sex chromosomes (Bachtrog 2006; Disteche 2016). The major events included the loss of most Y genes over millions of years, evolution of dosage compensation mechanisms that silenced one large X chromosome in each somatic mammalian XX cell, the sculpting of the gene content of the X and Y chromosomes, and evolution of other balancing mechanisms to

 Cite this article as *Cold Spring Harb Perspect Biol* doi: 10.1101/cshperspect.a039057

compensate for the evolving inequality of X and Y chromosomes (Vallender and Lahn 2004; Graves 2006; Hughes and Page 2015).

The loss of most Y genes is explained because the region near *Sry* was inherited only by males, so that it was insulated from any selection pressure to benefit females. The region thus diverged from the formerly homologous region of the evolving X chromosome and acquired mutations that benefitted males. The near-Sry region ceased to recombine with the X chromosome and entered a separate evolutionary path from the X chromosome, with different selection pressures and with progressively less recombination between the two chromosomes. This male-specific portion of the Y chromosome expanded over evolutionary time, and eventually included most of the Y chromosome. Nonrecombining (i.e., asexually inherited) DNA accumulates deleterious mutations (Charlesworth and Charlesworth 2000; Vallender and Lahn 2004; Bachtrog 2006), and thus most Y genes were lost. The genes remaining on the Y chromosome survived either because of their importance for males (for example, *Sry* and genes involved in testis functions such as spermatogenesis) or because they were involved in essential functions in many cell types and thus a reduction in their dosage was not tolerated (Bellott et al. 2014). The loss of Y genes meant that the homologous X genes were expressed higher in XX cells than XY cells because of the different numbers of the X chromosome, which triggered dosage-compensation mechanisms to offset the imbalance of gene expression (Julien et al. 2012). The main sex-specific dosage-compensation mechanism in mammals is mediated by the noncoding gene *Xist*, an X-linked gene that is expressed from one X chromosome in XX but not XY cells (Disteche 2016). *Xist* causes transcriptional silencing in *cis* of one X chromosome in XX cells, which effectively makes XX and XY cells more comparable in their expression of X genes. Thus, sexual differentiation of the X chromosome led to greater male–female equality.

Note that the emergence of major sex differences in the genome (degeneration of the Y chromosome and dosage compensation of X genes) are not explained because they enhance

reproduction. Of course, each of these changes must have occurred within the constraints of successful reproduction, but they apparently did not solve a problem of insufficient reproductive capacity of one or both sexes. Rather, most differences in the X and Y chromosome, and therefore sex differences in XX and XY cells, evolved because of unavoidable genomic forces that stem from asexual inheritance of most of the Y chromosome, which is passed from father to son without recombination. The number and type of sex chromosomes have distinct effects to cause sex differences in categories of X and Y genes, which then differentially regulate gene networks predominantly composed of autosomal genes (Raznahan et al. 2018).

SOME SEX DIFFERENCES EVOLVED TO REDUCE AND COMPENSATE FOR OTHER SEX DIFFERENCES

A major tenet of sex chromosome evolution theory is that sex differences in many gene pathways are maladaptive, and that natural selection operated to reduce or minimize sex differences in gene expression. Both males and females have equal need for cellular processes that support basic cellular functions (energy utilization, cell division, synthesis of RNA and proteins, etc.) not involved in reproduction per se. The genes on the autosomal chromosomal pair that evolved to become the X and Y chromosomes were involved in many gene pathways, always interacting with the much more numerous autosomal genes to support basic cellular processes that were required equally in the two sexes. As XX cells progressively began to express different levels of X gene products, relative to XY cells, basic cellular processes could not be equally optimal in XX and XY cells. Thus, there arose selection pressures for a compensatory mechanism. Selection pressures in males may have favored up-regulation of X genes when their former alleles were lost on the Y chromosome to make X genes more equally expressed in XY cells compared to XX cells. But these higher-expressing X alleles would in turn have been maladaptive when passed to XX cells because they would have caused too-high expression. These problems presumably created selection pressures to reduce expression of X genes only in XX females. In mammals, these pressures led to emergence of *Xist*-mediated reduction of X gene expression only in XX females. The molecular mechanisms of dosage compensation of X genes differs across species with XX-XY sex chromosomes (Lucchesi 2018), reinforcing the idea that selection pressures in favor of sexual equality (sex chromosome dosage compensation) were strong and potentially ubiquitous among species with heteromorphic sex chromosomes (Arnold et al. 2008). However, compensation was achieved by a variety of different molecular mechanisms in different taxa. In each case, however, the dosage-compensation mechanism must be sexually differentiated because of the requirement to adjust the dose of X genes in one sex relative to the other. In mammals, both XX and XY cells express X genes from one chromosome, which has also favored a second dosage-balancing mechanism that increased X expression in both sexes to be comparable to that of autosomal genes with which the X genes interact in gene networks (Disteche 2016).

The forces leading to sexual differentiation of the sex chromosomes were generally unrelated to reproduction. Despite the inherent involvement of *Sry* itself in male reproduction, and the accumulation of testis genes on the Y chromosome, numerous other sex-specific changes in the sex chromosome were driven by forces of genome evolution and intragenomic competition, not by selection pressures favoring better reproduction. The loss of Y genes is the result of properties of asexually inherited DNA. The male-specific expression of Y genes evolved not necessarily as an improvement on mechanisms of male reproduction, but as a by-product of the loss of recombination of X and Y chromosomes resulting from divergence of Y and X DNA. The sex-specific nature of *Xist* expression and dosage compensation is seen not as an adaption for better reproduction, but as a sex-specific process that offsets maladaptive sex differences caused by the evolving sexual inequality of X gene dose as the Y chromosome lost genes. In other words, sex differences in effects of any sex-biasing variable (sex hormones, sex chromosome genes) may evolve not because they mitigate problems of reproduction, but to offset a sex difference that has nothing to do with different reproductive roles but rather is a by-product of some other unavoidable evolutionary process. Returning to the issue of sex differences in metabolism, an alternative idea is that sex differences in sex steroid effects in metabolic tissues might also evolve as a mechanism to offset other sex differences that are maladaptive.

The concept that some sex differences are maladaptive is particularly intriguing because of the salience of this idea in theories of sex chromosome evolution. If some sex differences are caused by an inherent imbalance of X and Y gene effects in XX and XY cells, then the interactions of sex steroids and sex chromosome effects may also have nothing to do with reproduction per se. For example, XX inbred mice have greater body weight and adiposity than XY mice if one controls for levels of gonadal hormones (Chen et al. 2012). This sex chromosome effect is explained in part by the inherently higher expression of *Kdm5c*, an X gene that escapes inactivation and is expressed from all X chromosomes (Link et al. 2020). However, estrogens from the ovaries counteract the greater adiposity found in XX mice, and in general seem to blunt the effects of *Kdm5c* (Chen et al. 2012). Has the female-specific estrogen effect to reduce body fat evolved because it solves a problem of reproduction, or does it reduce the unequal effects of X genes in XX and XY cells? Are the sex-biased effects of gonadal hormones related to reproduction or maintenance of better sexual equality in the face of sex-biased genomic evolution? Or both?

The list of the major sex-biasing factors includes a variety of gonadal hormones such as testosterone, estradiol, and progesterone, which can also be synthesized in other tissues. The hormone effects can be organizational and/or activational. In parallel, various X and Y genes are now being discovered that cause sex differences in physiology and disease. All of these agents have pleiotropic, diverse effects on many downstream pathways in different tissues, which means that some of the effects may be maladaptive. Thus, some sex differences caused by one factor may be selected for if they offset a maladaptive sex difference caused by another factor.

For example, local synthesis of a sex hormone (an activational effect) might compensate for a maladaptive sex difference caused by organizational sexually differentiating effects of the same gonadal hormone. A related idea is that two different sources of sex steroids can cause offsetting effects that cancel each other. For example, local synthesis of estradiol in the neonatal female rat hippocampus may eliminate sex differences in hippocampal development caused by testicular secretion of testosterone, which is converted to estradiol in the male's brain (Amateau et al. 2004; Kight and McCarthy 2020).

SIGNATURES OF NONGONADAL SEXUAL DIMORPHISM ON THE X CHROMOSOME

Recent mapping of X chromosome genes showing sex differences in expression reveals a spatial distribution of such genes along the X chromosome that is better explained by the evolution of sex chromosomes than by a dominant effect of gonadal hormones on sex differences (Tukiainen et al. 2017; Oliva et al. 2020). The X chromosome can be divided into two segments. The pseudoautosomal region (PAR) was named by Paul Burgoyne because this region acts like an autosome (Burgoyne 1982). The XPAR and YPAR are homologous, and are the site of pairing of X and Y chromosome during meiosis. The other segment is the non-PAR (NPAR) X chromosome (NPX), which contains the vast majority of X genes in mammals. The NPX is the site of action of *Xist*, which transcriptionally silences most of the genes in one NPX in each XX cell. A major issue for sexual differentiation is that some X genes escape complete inactivation and are expressed from the "inactive" NPX chromosome in XX cells (Berletch et al. 2010). These X inactivation "escapees" are expressed constitutively higher in XX cells than XY cells, and thus represent an inherent source of sex bias in cell functions. In humans, up to 25% of NPX genes are expressed higher in XX than XY cells, although the degree of sex difference in expression varies across tissue type and developmental stage (Carrel and Willard 2005; Tukiainen et al. 2017). In mice, only about 3%–8% of X genes are X escapees (Berletch et al. 2015). In experiments that manipulate the dose of X genes while measuring effects on disease processes, X escapees are shown to contribute to sex differences in a variety of traits, including effects of prenatal stress (Nugent et al. 2018), bladder cancer (Kaneko and Li 2018), autoimmune disease (Itoh et al. 2019), metabolism and adiposity (Link et al. 2020), Alzheimer's disease (Davis et al. 2020), and stroke (Qi et al. 2021).

In contrast to X escapees from NPX, PAR genes have long been thought to not be a source of sex-biasing factors. Because both XX and XY cells have two copies of the PAR, no sex differences in expression of PAR genes were expected. However, well-powered studies of gene expression in human tissues demonstrate that many PAR genes are expressed higher in XY than XX cells (Tukiainen et al. 2017). It appears that the effects of *Xist* spill over into the PAR, so that expression in XX cells is reduced from the inactive X chromosome, whereas *Xist* is not expressed in XY cells so that PAR genes are more fully expressed from both sex chromosomes. Thus, we expect PAR genes to be a class of X genes causing sexual bias, and they should be a focus for future studies.

Figure 3 shows a pattern of sexual dimorphism of gene expression across the human X chromosome. We see that genes expressed higher in females (XX) are concentrated in the NPAR, whereas genes expressed higher in males (XY) are concentrated in the PAR. A priori, these sex differences in gene expression could be caused either by gonadal hormone effects, or by sex chromosome effects related to X dosage controlled by X inactivation and escape. In fact, these sex differences are likely caused by both classes of factors. However, the pattern of direction of sex bias (male bias in PAR, female bias in NPAR) is explained by sex chromosome dosage mechanisms, not by hormonal influences. There is no basis for expecting that gonadal hormones would have this specific pattern.

IMPLICATIONS FOR EDUCATION IN THE FIELD OF SEXUAL DIFFERENTIATION

The number of sex chromosome effects reported is still vastly smaller than the number of gonadal

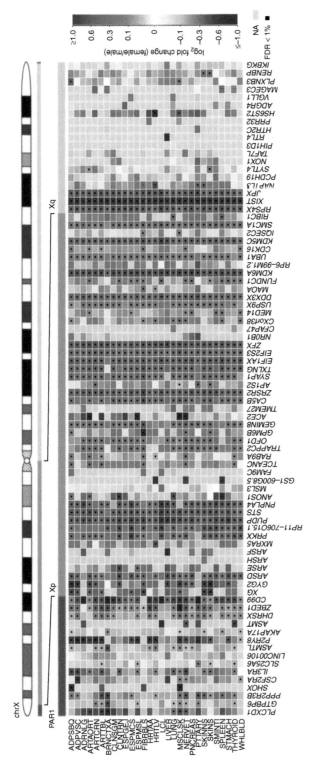

Figure 3. Sex differences in expression of human X chromosome genes. The figure summarizes sex differences in expression of human X chromosome genes in 29 different tissues (rows) mapped by relative position along the X chromosome (columns). Data are derived from analysis of 5000 transcriptomes from Genotype-Tissue Expression (GTEx) consortium data. Red indicates expression higher in females, and blue indicates expression higher in males. Gray indicates expression too low to be analyzed. The non-pseudoautosomal region (NPAR) contains about 23% of X genes with overall higher expression in females, in contrast to the PAR1 region where most genes were expressed higher in males. Although the sex differences in expression could be caused by any sex-biasing agent, the general spatial pattern of sex bias is better predicted by patterns of X inactivation and escape from inactivation by forces of sex chromosome evolution rather than a hormonal theory. (Reprinted from Tukiainen et al. 2017 under the terms of the Creative Commons Attribution 4.0 International license.)

hormonal effects. Hormonal effects have been studied for at least a century in many species, using basic methods of gonadectomy and endocrine replacement. In contrast, sex chromosome effects are best studied so far only in mice, using unusual models that have come into use in the last 20 years (Arnold 2020). Most phenotypes affected by sex-biasing sex chromosome effects are also sexually differentiated by effects of gonadal hormones. As discussed above, sex chromosome effects have been reported to help explain sex differences in bladder cancer (Kaneko and Li 2018), autoimmune disease such as multiple sclerosis (Voskuhl and Gold 2012), adiposity (Link and Reue 2017), stroke (McCullough et al. 2016), cardiovascular diseases (Arnold et al. 2017), and Alzheimer's disease (Davis et al. 2020). In each of these cases, gonadal hormones also influence disease progression in mouse models. Yet, there has been almost no published research addressing how the two types of effects interact at the molecular and cellular level. In part, this omission may stem from the relative isolation of the investigators studying endocrine versus sex chromosome effects. Education in the field of sexual differentiation may just now be incorporating new ideas suggesting that sex chromosome effects are important and interact with sex hormone effects. As discussed above, the unequal effects of X and Y genes did not evolve primarily as adaptations promoting female or male forms of reproduction, and therefore may be outside of the conceptual framework of the majority of scientists studying sexual differentiation. Future investigators of sexual differentiation may benefit from an appreciation not only of the evolution of endocrine control of sexual dimorphism, but also the evolution of sexual imbalance and balance in the sex chromosomes. The combined understanding of divergent classes of sex-biasing factors will be required to understand their interaction in the complex control of sex differences in physiology and disease.

SUMMARY

We have summarized two conceptual frameworks that are used to explain the evolution of sex differences in phenotypes. The classic view is that the inherent difference in levels of gonadal hormones causes sexually dimorphic development of tissues and phenotypes that are adapted for different roles in reproduction. The view from studies of sex chromosomes indicates that sex differences in the dose of sex chromosome genes evolved because of intrinsic genomic forces that were not primarily adaptations for improved reproductive performance. Thus, sex differences in gene expression evolved that are disadvantageous in one or both sexes in many cases, leading to counteracting sex-specific selection pressures to rebalance gene expression between the two sexes. When sexual equivalence exists, it may be achieved by separate mechanisms in the two sexes. The evolution of sexual dimorphism on the sex chromosomes is thus a continual process of unbalancing and rebalancing gene expression. When a disease process influences part of the rebalancing mechanism (for example, X-inactivation in females but not males), it can disrupt the balance and affect one sex more than the other, leading to sex differences in prognosis or progression of the disease. These ideas suggest that improved understanding of sex differences in physiology and disease will be improved by investigating both gonadal hormone and sex chromosome effects and their interaction. The next generation of sexual differentiationists will have a stronger foundation if they learn lessons from diverse hormonal and sex chromosomal investigations.

ACKNOWLEDGMENTS

Many thanks to members of my laboratory (especially Xuqi Chen and Yuichiro Itoh), to collaborators at UCLA (Karen Reue, Rhonda Voksuhl, Mansoureh Eghbali, Jake Lusis, Xia Yang, Eric Vilain, and Allan Mackenzie-Graham), and members of their laboratories, and to many colleagues elsewhere (especially Paul Burgoyne, Jenny Graves, Andrew Sinclair), for innumerable conversations about the issues discussed here. The author's research has been supported by NIH grants HD100298, OD026560, HL131182, HD076125, and DK083561.

REFERENCES

Achermann JC, Jameson JL. 2017. Disorders of sex development. In *Harrison's endocrinology* (ed. Jameson JL), pp. 146–158. McGraw Hill, New York.

Alsiraj Y, Thatcher SE, Charnigo R, Chen K, Blalock E, Daugherty A, Cassis LA. 2017. Female mice with an XY sex chromosome complement develop severe angiotensin II–induced abdominal aortic aneurysms. *Circulation* **135:** 379–391. doi:10.1161/CIRCULATIONAHA.116.023789

Alsiraj Y, Thatcher SE, Blalock E, Fleenor B, Daugherty A, Cassis LA. 2018. Sex chromosome complement defines diffuse versus focal angiotensin II–induced aortic pathology. *Arterioscler Thromb Vasc Biol* **38:** 143–153. doi:10.1161/ATVBAHA.117.310035

Alsiraj Y, Chen X, Thatcher SE, Temel RE, Cai L, Blalock E, Katz W, Ali HM, Petriello M, Deng P, et al. 2019. XX sex chromosome complement promotes atherosclerosis in mice. *Nat Commun* **10:** 2631. doi:10.1038/s41467-019-10462-z

Amateau SK, Alt JJ, Stamps CL, McCarthy MM. 2004. Brain estradiol content in newborn rats: sex differences, regional heterogeneity, and possible de novo synthesis by the female telencephalon. *Endocrinology* **145:** 2906–2917. doi:10.1210/en.2003-1363

Arnold AP. 2002. Concepts of genetic and hormonal induction of vertebrate sexual differentiation in the twentieth century, with special reference to the brain. In *Hormones, brain, and behavior* (ed. Pfaff DW, Arnold AP, Etgen A, Fahrbach S, Rubin R), pp. 105–135. Academic Press, San Diego.

Arnold AP. 2009. The organizational-activational hypothesis as the foundation for a unified theory of sexual differentiation of all mammalian tissues. *Horm Behav* **55:** 570–578. doi:10.1016/j.yhbeh.2009.03.011

Arnold AP. 2019. Rethinking sex determination of non-gonadal tissues. *Curr Top Dev Biol* **134:** 289–315. doi:10.1016/bs.ctdb.2019.01.003

Arnold AP. 2020. Four core genotypes and XY* mouse models: update on impact on SABV research. *Neurosci Biobehav Rev* **119:** 1–8. doi:10.1016/j.neubiorev.2020.09.021

Arnold AP, Itoh Y, Melamed E. 2008. A bird's-eye view of sex chromosome dosage compensation. *Annu Rev Genomics Hum Genet* **9:** 109–127. doi:10.1146/annurev.genom.9.081307.164220

Arnold AP, Chen X, Link JC, Itoh Y, Reue K. 2013. Cell-autonomous sex determination outside of the gonad. *Dev Dyn* **242:** 371–379. doi:10.1002/dvdy.23936

Arnold AP, Cassis LA, Eghbali M, Reue K, Sandberg K. 2017. Sex hormones and sex chromosomes cause sex differences in the development of cardiovascular diseases. *Arterioscler Thromb Vasc Biol* **37:** 746–756. doi:10.1161/ATVBAHA.116.307301

Bachtrog D. 2006. A dynamic view of sex chromosome evolution. *Curr Opin Genet Dev* **16:** 578–585. doi:10.1016/j.gde.2006.10.007

Bakker J, Brock O. 2010. Early oestrogens in shaping reproductive networks: evidence for a potential organisational role of oestradiol in female brain development. *J Neuroendocrinol* **22:** 728–735.

Barker JM, Torregrossa MM, Arnold AP, Taylor JR. 2010. Dissociation of genetic and hormonal influences on sex differences in alcoholism-related behaviors. *J Neurosci* **30:** 9140–9144. doi:10.1523/JNEUROSCI.0548-10.2010

Bellott DW, Hughes JF, Skaletsky H, Brown LG, Pyntikova T, Cho TJ, Koutseva N, Zaghlul S, Graves T, Rock S, et al. 2014. Mammalian Y chromosomes retain widely expressed dosage-sensitive regulators. *Nature* **508:** 494–499. doi:10.1038/nature13206

Berletch JB, Yang F, Disteche CM. 2010. Escape from X inactivation in mice and humans. *Genome Biol* **11:** 213. doi:10.1186/gb-2010-11-6-213

Berletch JB, Ma W, Yang F, Shendure J, Noble WS, Disteche CM, Deng X. 2015. Escape from X inactivation varies in mouse tissues. *PLoS Genet* **11:** e1005079. doi:10.1371/journal.pgen.1005079

Bermejo-Alvarez P, Rizos D, Lonergan P, Gutierrez-Adan A. 2011. Transcriptional sexual dimorphism during preimplantation embryo development and its consequences for developmental competence and adult health and disease. *Reproduction* **141:** 563–570. doi:10.1530/REP-10-0482

Bonthuis PJ, Cox KH, Rissman EF. 2012. X-chromosome dosage affects male sexual behavior. *Horm Behav* **61:** 565–572. doi:10.1016/j.yhbeh.2012.02.003

Burgoyne PS. 1982. Genetic homology and crossing over in the X and Y chromosomes of mammals. *Hum Genet* **61:** 85–90. doi:10.1007/BF00274192

Burgoyne PS, Arnold AP. 2016. A primer on the use of mouse models for identifying direct sex chromosome effects that cause sex differences in non-gonadal tissues. *Biol Sex Differ* **7:** 68. doi:10.1186/s13293-016-0115-5

Capel B. 2017. Vertebrate sex determination: evolutionary plasticity of a fundamental switch. *Nat Rev Genet* **18:** 675–689. doi:10.1038/nrg.2017.60

Carrel L, Willard HF. 2005. X-inactivation profile reveals extensive variability in X-linked gene expression in females. *Nature* **434:** 400–404. doi:10.1038/nature03479

Carruth LL, Reisert I, Arnold AP. 2002. Sex chromosome genes directly affect brain sexual differentiation. *Nat Neurosci* **5:** 933–934. doi:10.1038/nn922

Case LK, Wall EH, Dragon JA, Saligrama N, Krementsov DN, Moussawi M, Zachary JF, Huber SA, Blankenhorn EP, Teuscher C. 2013. The Y chromosome as a regulatory element shaping immune cell transcriptomes and susceptibility to autoimmune disease. *Genome Res* **23:** 1474–1485.

Charlesworth B. 2003. The organization and evolution of the human Y chromosome. *Genome Biol* **4:** 226. doi:10.1186/gb-2003-4-9-226

Charlesworth B, Charlesworth D. 2000. The degeneration of Y chromosomes. *Philos Trans R Soc Lond B Biol Sci* **355:** 1563–1572. doi:10.1098/rstb.2000.0717

Charlesworth D, Charlesworth B, Marais G. 2005. Steps in the evolution of heteromorphic sex chromosomes. *Heredity (Edinb)* **95:** 118–128. doi:10.1038/sj.hdy.6800697

Chen X, Watkins R, Delot E, Reliene R, Schiestl RH, Burgoyne PS, Arnold AP. 2008. Sex difference in neural tube defects in p53-null mice is caused by differences in the complement of X not Y genes. *Dev Neurobiol* **68:** 265–273. doi:10.1002/dneu.20581

Cite this article as *Cold Spring Harb Perspect Biol* doi: 10.1101/cshperspect.a039057

Chen X, Grisham W, Arnold AP. 2009. X chromosome number causes sex differences in gene expression in adult mouse striatum. *Eur J Neurosci* **29:** 768–776. doi:10.1111/j.1460-9568.2009.06610.x

Chen X, McClusky R, Chen J, Beaven SW, Tontonoz P, Arnold AP, Reue K. 2012. The number of X chromosomes causes sex differences in adiposity in mice. *PLoS Genet* **8:** e1002709. doi:10.1371/journal.pgen.1002709

Chen X, McClusky R, Itoh Y, Reue K, Arnold AP. 2013. X and Y chromosome complement influence adiposity and metabolism in mice. *Endocrinology* **154:** 1092–1104. doi:10.1210/en.2012-2098

Chen X, Wang L, Loh DH, Colwell CS, Taché Y, Reue K, Arnold AP. 2015. Sex differences in diurnal rhythms of food intake in mice caused by gonadal hormones and complement of sex chromosomes. *Horm Behav* **75:** 55–63. doi:10.1016/j.yhbeh.2015.07.020

Cooke B, Hegstrom CD, Villeneuve LS, Breedlove SM. 1998. Sexual differentiation of the vertebrate brain: principles and mechanisms. *Front Neuroendocrinol* **19:** 323–362. doi:10.1006/frne.1998.0171

Corre C, Friedel M, Vousden DA, Metcalf A, Spring S, Qiu LR, Lerch JP, Palmert MR. 2016. Separate effects of sex hormones and sex chromosomes on brain structure and function revealed by high-resolution magnetic resonance imaging and spatial navigation assessment of the four core genotype mouse model. *Brain Struct Funct* **221:** 997–1016. doi:10.1007/s00429-014-0952-0

Cox KH, Rissman EF. 2011. Sex differences in juvenile mouse social behavior are influenced by sex chromosomes and social context. *Genes Brain Behav* **10:** 465–472. doi:10.1111/j.1601-183X.2011.00688.x

Cox KH, Bonthuis PJ, Rissman EF. 2014. Mouse model systems to study sex chromosome genes and behavior: relevance to humans. *Front Neuroendocrinol* **35:** 405–419. doi:10.1016/j.yfrne.2013.12.004

Cox KH, Quinnies KM, Eschendroeder A, Didrick PM, Eugster EA, Rissman EF. 2015. Number of X-chromosome genes influences social behavior and vasopressin gene expression in mice. *Psychoneuroendocrinology* **51:** 271–281. doi:10.1016/j.psyneuen.2014.10.010

Davis EJ, Broestl L, Abdulai-Saiku S, Worden K, Bonham LW, Miñones-Moyano E, Moreno AJ, Wang D, Chang K, Williams G, et al. 2020. A second X chromosome contributes to resilience in a mouse model of Alzheimer's disease. *Sci Transl Med* **12:** eaaz5677. doi:10.1126/scitranslmed.aaz5677

Delbridge ARD, Kueh AJ, Ke F, Zamudio NM, El-Saafin F, Jansz N, Wang GY, Iminitoff M, Beck T, Haupt S, et al. 2019. Loss of p53 causes stochastic aberrant X-chromosome inactivation and female-specific neural tube defects. *Cell Rep* **27:** 442–454.e5. doi:10.1016/j.celrep.2019.03.048

De Vries GJ. 2004. Minireview: sex differences in adult and developing brains: compensation, compensation, compensation. *Endocrinology* **145:** 1063–1068. doi:10.1210/en.2003-1504

De Vries GJ, Rissman EF, Simerly RB, Yang LY, Scordalakes EM, Auger CJ, Swain A, Lovell-Badge R, Burgoyne PS, Arnold AP. 2002. A model system for study of sex chromosome effects on sexually dimorphic neural and behavioral traits. *J Neurosci* **22:** 9005–9014. doi:10.1523/JNEUROSCI.22-20-09005.2002

Disteche CM. 2016. Dosage compensation of the sex chromosomes and autosomes. *Semin Cell Dev Biol* **56:** 9–18. doi:10.1016/j.semcdb.2016.04.013

Du S, Itoh N, Askarinam S, Hill H, Arnold AP, Voskuhl RR. 2014. XY sex chromosome complement, compared with XX, in the CNS confers greater neurodegeneration during experimental autoimmune encephalomyelitis. *Proc Natl Acad Sci* **111:** 2806–2811. doi:10.1073/pnas.1307091111

Gioiosa L, Chen X, Watkins R, Klanfer N, Bryant CD, Evans CJ, Arnold AP. 2008. Sex chromosome complement affects nociception in tests of acute and chronic exposure to morphine in mice. *Horm Behav* **53:** 124–130. doi:10.1016/j.yhbeh.2007.09.003

Graves JAM. 2006. Sex chromosome specialization and degeneration in mammals. *Cell* **124:** 901–914. doi:10.1016/j.cell.2006.02.024

Hughes JF, Page DC. 2015. The biology and evolution of mammalian Y chromosomes. *Annu Rev Genet* **49:** 507–527. doi:10.1146/annurev-genet-112414-055311

Itoh Y, Golden LC, Itoh N, Matsukawa MA, Ren E, Tse V, Arnold AP, Voskuhl RR. 2019. The X-linked histone demethylase Kdm6a in CD4[+] T lymphocytes modulates autoimmunity. *J Clin Invest* **129:** 3852–3863. doi:10.1172/JCI126250

Ji H, Zheng W, Wu X, Liu J, Ecelbarger CM, Watkins R, Arnold AP, Sandberg K. 2010. Sex chromosome effects unmasked in angiotensin II–induced hypertension. *Hypertension* **55:** 1275–1282. doi:10.1161/HYPERTENSIONAHA.109.144949

Jost A. 1970. Hormonal factors in the sex differentiation of the mammalian foetus. *Phil Trans Roy Soc Lond B Biol Sci* **259:** 119–131. doi:10.1098/rstb.1970.0052

Julien P, Brawand D, Soumillon M, Necsulea A, Liechti A, Schütz F, Daish T, Grützner F, Kaessmann H. 2012. Mechanisms and evolutionary patterns of mammalian and avian dosage compensation. *PLoS Biol* **10:** e1001328. doi:10.1371/journal.pbio.1001328

Kaneko S, Li X. 2018. X chromosome protects against bladder cancer in females via a *KDM6A*-dependent epigenetic mechanism. *Sci Adv* **4:** eaar5598. doi:10.1126/sciadv.aar5598

Kight KE, McCarthy MM. 2020. Androgens and the developing hippocampus. *Biol Sex Differ* **11:** 30. doi:10.1186/s13293-020-00307-6

Kuljis DA, Loh DH, Truong D, Vosko AM, Ong ML, McClusky R, Arnold AP, Colwell CS. 2013. Gonadal- and sex-chromosome-dependent sex differences in the circadian system. *Endocrinology* **154:** 1501–1512. doi:10.1210/en.2012-1921

Li J, Chen X, McClusky R, Ruiz-Sundstrom M, Itoh Y, Umar S, Arnold AP, Eghbali M. 2014. The number of X chromosomes influences protection from cardiac ischaemia/reperfusion injury in mice: one X is better than two. *Cardiovasc Res* **102:** 375–384. doi:10.1093/cvr/cvu064

Lillie FR. 1939. General biological introduction. In *Sex and internal secretions* (ed. Allen E, Danforth CH, Doisy EA), pp. 3–14. Williams and Wilkins, Baltimore.

Link JC, Reue K. 2017. Genetic basis for sex differences in obesity and lipid metabolism. *Annu Rev Nutr* **37:** 225–245. doi:10.1146/annurev-nutr-071816-064827

Link JC, Chen X, Prien C, Borja MS, Hammerson B, Oda MN, Arnold AP, Reue K. 2015. Increased high-density lipoprotein cholesterol levels in mice with XX versus XY sex chromosomes. *Arterioscler Thromb Vasc Biol* **35:** 1778–1786. doi:10.1161/ATVBAHA.115.305460

Link JC, Hasin-Brumshtein Y, Cantor RM, Chen X, Arnold AP, Lusis AJ, Reue K. 2017. Diet, gonadal sex, and sex chromosome complement influence white adipose tissue miRNA expression. *BMC Genomics* **18:** 89. doi:10.1186/s12864-017-3484-1

Link JC, Wiese CB, Chen X, Avetisyan R, Ronquillo E, Ma F, Guo X, Yao J, Allison M, Chen YDI, et al. 2020. X chromosome dosage of histone demethylase KDM5C determines sex differences in adiposity. *J Clin Invest* **130:** 5688–5702. doi:10.1172/JCI140223

Lowe R, Gemma C, Rakyan VK, Holland ML. 2015. Sexually dimorphic gene expression emerges with embryonic genome activation and is dynamic throughout development. *BMC Genomics* **16:** 295. doi:10.1186/s12864-015-1506-4

Lucchesi JC. 2018. Transcriptional modulation of entire chromosomes: dosage compensation. *J Genet* **97:** 357–364. doi:10.1007/s12041-018-0919-7

Maggi A, Della Torre S. 2018. Sex, metabolism and health. *Mol Metab* **15:** 3–7. doi:10.1016/j.molmet.2018.02.012

McCarthy MM, Arnold AP. 2011. Reframing sexual differentiation of the brain. *Nat Neurosci* **14:** 677–683. doi:10.1038/nn.2834

McCullough LD, Mirza MA, Xu Y, Bentivegna K, Steffens EB, Ritzel R, Liu F. 2016. Stroke sensitivity in the aged: sex chromosome complement vs. gonadal hormones. *Aging (Albany NY)* **8:** 1432–1441. doi:10.18632/aging.100997

Midgeon BR. 2007. *Females are mosaics.* Oxford University Press, Oxford.

Nugent BM, O'Donnell CM, Epperson CN, Bale TL. 2018. Placental H3K27me3 establishes female resilience to prenatal insults. *Nat Commun* **9:** 2555. doi:10.1038/s41467-018-04992-1

Oliva M, Muñoz-Aguirre M, Kim-Hellmuth S, Wucher V, Gewirtz ADH, Cotter DJ, Parsana P, Kasela S, Balliu B, Viñuela A, et al. 2020. The impact of sex on gene expression across human tissues. *Science* **369:** eaba3066. doi:10.1126/science.aba3066

Palaszynski KM, Smith DL, Kamrava S, Burgoyne PS, Arnold AP, Voskuhl RR. 2005. A yin-yang effect between sex chromosome complement and sex hormones on the immune response. *Endocrinology* **146:** 3280–3285. doi:10.1210/en.2005-0284

Phoenix CH, Goy RW, Gerall AA, Young WC. 1959. Organizing action of prenatally administered testosterone propionate on the tissues mediating mating behavior in the female Guinea pig. *Endocrinology* **65:** 369–382. doi:10.1210/endo-65-3-369

Qi S, Al Mamun A, Ngwa C, Romana S, Ritzel R, Arnold AP, McCullough LD, Liu F. 2021. X chromosome escapee genes are involved in ischemic sexual dimorphism through epigenetic modification of inflammatory signals. *J Neuroinflammation* **18:** 70. doi:10.1186/s12974-021-02120-3

Quinn JJ, Hitchcott PK, Umeda EA, Arnold AP, Taylor JR. 2007. Sex chromosome complement regulates habit formation. *Nat Neurosci* **10:** 1398–1400. doi:10.1038/nn1994

Raznahan A, Parikshak NN, Chandran V, Blumenthal JD, Clasen LS, Alexander-Bloch AF, Zinn AR, Wangsa D, Wise J, Murphy DGM, et al. 2018. Sex-chromosome dosage effects on gene expression in humans. *Proc Natl Acad Sci* **115:** 7398–7403. doi:10.1073/pnas.1802889115

Renfree MB, Short RV. 1988. Sex determination in marsupials: evidence for a marsupial—eutherian dichotomy. *Philos Trans R Soc Lond B Biol Sci* **322:** 41–53. doi:10.1098/rstb.1988.0112

Shi W, Sheng X, Dorr KM, Hutton JE, Emerson JI, Davies HA, Andrade TD, Wasson LK, Greco TM, Hashimoto Y, et al. 2021. Cardiac proteomics reveals sex chromosome-dependent differences between males and females that arise prior to gonad formation. *Dev Cell* **56:** 3019–3034.e7. doi:10.1016/j.devcel.2021.09.022

Smith-Bouvier DL, Divekar AA, Sasidhar M, Du S, Tiwari-Woodruff SK, King JK, Arnold AP, Singh RR, Voskuhl RR. 2008. A role for sex chromosome complement in the female bias in autoimmune disease. *J Exp Med* **205:** 1099–1108. doi:10.1084/jem.20070850

Sutton E, Hughes J, White S, Sekido R, Tan J, Arboleda V, Rogers N, Knower K, Rowley L, Eyre H, et al. 2011. Identification of SOX3 as an XX male sex reversal gene in mice and humans. *J Clin Invest* **121:** 328–341. doi:10.1172/JCI42580

Taylor AMW, Chadwick CI, Mehrabani S, Hrncir H, Arnold AP, Evans CJ. 2022. Sex differences in κ opioid receptor antinociception is influenced by the number of X chromosomes in mouse. *J Neurosci Res* **100:** 183–190. doi:10.1002/jnr.24704

Tukiainen T, Villani AC, Yen A, Rivas MA, Marshall JL, Satija R, Aguirre M, Gauthier L, Fleharty M, Kirby A, et al. 2017. Landscape of X chromosome inactivation across human tissues. *Nature* **550:** 244–248. doi:10.1038/nature24265

Vallender EJ, Lahn BT. 2004. How mammalian sex chromosomes acquired their peculiar gene content. *Bioessays* **26:** 159–169. doi:10.1002/bies.10393

Voskuhl RR, Gold SM. 2012. Sex-related factors in multiple sclerosis susceptibility and progression. *Nat Rev Neurol* **8:** 255–263. doi:10.1038/nrneurol.2012.43

Vousden DA, Corre C, Spring S, Qiu LR, Metcalf A, Cox E, Lerch JP, Palmert MR. 2018. Impact of X/Y genes and sex hormones on mouse neuroanatomy. *Neuroimage* **173:** 551–563. doi:10.1016/j.neuroimage.2018.02.051

Yildirim E, Kirby JE, Brown DE, Mercier FE, Sadreyev RI, Scadden DT, Lee JT. 2013. *Xist* RNA is a potent suppressor of hematologic cancer in mice. *Cell* **152:** 727–742. doi:10.1016/j.cell.2013.01.034

Yu B, Qi Y, Li R, Shi Q, Satpathy AT, Chang HY. 2021. B cell-specific XIST complex enforces X-inactivation and restrains atypical B cells. *Cell* **184:** 1790–1803.e17. doi:10.1016/j.cell.2021.02.015

Cite this article as *Cold Spring Harb Perspect Biol* doi: 10.1101/cshperspect.a039057

Index